T0400765

PHARMACOLOGY - RESEARCH, SAFETY TESTING AND REGULATION

PROTON PUMP INHIBITORS (PPIS)

PREVALENCE OF USE, EFFECTIVENESS AND IMPLICATIONS FOR CLINICIANS

PHARMACOLOGY - RESEARCH, SAFETY TESTING AND REGULATION

Additional books in this series can be found on Nova's website under the Series tab.

Additional e-books in this series can be found on Nova's website under the e-book tab.

PHARMACOLOGY - RESEARCH, SAFETY TESTING AND REGULATION

PROTON PUMP INHIBITORS (PPIS)

PREVALENCE OF USE, EFFECTIVENESS AND IMPLICATIONS FOR CLINICIANS

BARBARA PARKER
EDITOR

New York

Library of Congress Cataloging-in-Publication Data

ISBN: 978-1-63482-890-1

Library of Congress Control Number: 2015939742

Published by Nova Science Publishers, Inc. † New York

Contents

Preface		**vii**
Chapter I	Proton Pump Inhibitors: Pharmacologic Principles of Use *Abhijeet Waghray, Nisheet Waghray, Birju Shah and M. Michael Wolfe*	**1**
Chapter II	Benefits and Risks of Routine Use of Proton Pump Inhibitors at the Intensive Care Unit *Lukas Buendgens and Frank Tacke*	**39**
Chapter III	Use and Misuse of Proton Pump Inhibitors: A Survey on a General Population *L. Lombardo, A. Giostra, L. Viganò, M. Tabone, A. Grassi, M. Gionco, G. Rovera, A. Bo, P. Malvasi and Appropriateness Study Mauriziano Hospital Group*	**51**
Chapter IV	Reported Complications of Proton Pump Inhibitor Use: An Update for the Clinician *Edward C. Oldfield IV and David A. Johnson*	**61**
Chapter V	Adverse Events of Long-Term Proton Pump Inhibitor Therapy *Nieka K. Harris, Amy N. Stratton and Fouad J. Moawad*	**125**
Index		**145**

Preface

Proton pump inhibitors (PPIs) has become a mainstay of therapy not only in the practice of gastroenterology, but also for general medicine and even for the general public. Recently, proton pump inhibitors have been associated with a number of potential complications including impaired vitamin and mineral absorption, altered drug metabolism and increased risk for infections. The outline for the basis for these reported risks are examined in this book, the current clinical evidence is discussed, and the authors offer insight into the likely clinical significance of each complication. Furthermore, the pathophysiology of acid secretion and inhibition, with a specific focus on the pharmacological principles surrounding the use of PPIs are reviewed. Relevant information on the management of acid-related disordered is provided for clinicians. Other chapters explore the current literature on potential complications associated with long-term PPI use, the benefits and risks of routine use of proton pump inhibitors in the intensive care unit (ICU), often associated with adverse events such as a significantly increased risk of infectious complications, and the use and misuse of proton pump inhibitors in the general population.

Chapter I – The secretion of large quantities of concentrated hydrochloric acid (HCl) by gastric parietal cells is a hallmark of the mammalian stomach. Since their development in the 1970s and 1980s, the mainstay of the antisecretory therapy had been histamine-2 receptor antagonists (H_2RAs). The recognition of the H^+/K^+-ATPase enzyme as the final pathway of HCl secretion led to the development of proton pump inhibitors (PPIs). PPIs are specific inhibitors of H^+/K^+-ATPase and function as prodrugs, requiring acid activation for maximum efficacy. When administered optimally, PPIs are the most potent inhibitors of gastric acid secretion, and their availability has expanded the therapeutic options available for acid-related disorders, including gastroduodenal ulcers, NSAID-related mucosal injury, Zollinger-Ellison syndrome, stress related ulcer syndrome in critically ill patients, gastroesophageal reflux disease, and *H. pylori* eradication therapy. As an increasing number of PPIs are being prescribed by physicians, less than optimal dosing has become an area of concern. The purpose of this chapter is to review the pathophysiology of acid secretion and inhibition, with a specific focus on the pharmacological principles surrounding the use of PPIs. Furthermore, relevant information on the management of acid-related disorders will be provided for clinicians.

Chapter II – Stress-related mucosal disease (SRMD) is present in nearly all critical ill patients. SRMD leads to an elevated risk of upper gastrointestinal (GI) bleedings, a potentially life-threatening complication associated with a mortality rate of about 30%. This led to the recommendation for stress ulcer prophylaxis in critically ill patients, especially to those with risk factors for GI bleeding such as mechanical ventilation or coagulopathies. Proton pump inhibitors (PPI) effectively prevent gastrointestinal bleedings in critically ill patients at the intensive care unit (ICU). Large meta-analyses including up to 1720 patients from 14 clinical trials revealed that PPI seem to be more effective than histamine 2 receptor antagonists (H2RA) in preventing clinically significant upper GI bleedings in critically ill patients, although they did not reduce overall mortality in ICU patients. Moreover, recent studies revealed that the routine use of PPI at the ICU can be associated with adverse events such as a significantly increased risk of infectious complications, especially of nosocomial pneumonia and *Clostridium difficile*–associated diarrhea (CDAD). Likewise, PPI can be toxic for both the liver and the bone marrow, and some PPI show clinically relevant interactions with important other drugs like clopidogrel. Therefore, the agent of choice, the specific balance of risks and benefits for individual patients as well as the possible dose of PPI has to be chosen carefully. Alternatives to PPI prophylaxis include histamine receptor blockers and/or sucralfate. Instead of routine PPI use for bleeding prophylaxis, further trials should investigate risk-adjusted algorithms, balancing benefits and threats of PPI medication at the ICU.

Chapter III – *Background:* Prescription rates of Proton pump inhibitors (PPI) continue to rise with growing concern over possible side effects and costs. *Aim*: To evaluate in real world: 1) the overall prevalence of short-term and long-term PPI use, 2) its relative appropriateness. *Materials & Methods*: From January to April 2011, all the out-patients asking for medical or surgical consultation at the Mauriziano U.I Hospital in Turin were investigated with a structured computerized pro-forma for detailed pharmacological history (dosage, prescribing and time modalities). Patients taking PPI for at least 6 months over the last 2 years were defined "long-term users" (LTU). Diagnostic procedures, final diagnosis, symptoms and drugs were registered. *Results:* Out of 1056 patients who entered the study (620 F, 436 M, median age 70 yrs, peak 75 yrs) 478 (306 F, 172 M; $p< 0.002$) were PPI users (45%). Among these, 373 were LTU (79%; F 238). Among LTU: 1) Prescription was made by the General Practitioner in 37%, by the Gastroenterologist in 30%, by other Specialists in 30% and as auto medication in 3% of the cases. 2) Diagnostic procedures: a previous UGIE was performed in 40%, oesophageal 24-h pHmetry in 2%, oesophageal manometry in 1% of the patients. Tests for Hp detection were performed in 45% of the patients. 3) Diagnosis was GERD in 31%, functional dyspepsia in 23%, unspecified gastritis in 7%, Hp-negative gastritis in 8%, Hp-positive gastritis in 10%, previous duodenal or gastric peptic ulcer in 5%, chronic atrophic gastritis in 4%, gastric neoplasia in 1%, diagnosis non specified in 11% of the cases. 4) Symptoms: when PPI treatment was discontinued pain/discomfort relapsed in 32%, continued to be absent in 38%, was not specified in 30%; at reintroduction of PPI pain/discomfort disappeared in 26%, persisted in 22%, partially regressed in 16%, was not specified in 36% of the patients. 5) Gastrolesive drugs were taken by 48% of patients. *Conclusions:* This study confirms that PPIs use in general population is extensive (45%) with a statistically significant prevalence in female patients ($p<0.002$), being of long-term type in the majority of the cases (79%). In LTU patients lack of appropriateness was observed for diagnostic procedures in 58% and for indications in 69%. It is worth noting that the diagnosis was "chronic atrophic gastritis" in 4% and not even specified in 11% of the cases. Clinicians should be aware of these realities in order to avoid side effects and unjustified social costs.

Chapter IV – Recently, proton pump inhibitors have been associated with a number of potential complications including impaired vitamin and mineral absorption, altered drug metabolism, and increased risk for infections. Subsequently, there has been a plethora of recent investigation into each of these areas, with conflicting results noted for the majority of the reported complications. This chapter aims to outline the basis for these reported risks, discuss the current clinical evidence, and also offer insight into the likely clinical significance of each complication.

Chapter V – Gastroesophageal reflux disease (GERD) is a common condition affecting approximately 10-30% of the Western population. Risk factors include obesity, diet, anatomic factors such as a hiatal hernia, and medications which may lower the lower esophageal sphincter (LES) resting pressure. With the rise in risk factors and subsequent increase in gastroesophageal reflux symptoms, proton pump inhibitors (PPI) are commonly prescribed and used long-term. Although PPIs are relatively safe and necessary when indicated, there are concerns regarding long-term adverse effects of chronic PPI use. With decreased gastric acid secretion, there exists a potential for micronutrient malabsorption, osteoporosis, and enteric infections. Less common complications of PPI usage include community associated pneumonia, interstitial nephritis and microscopic colitis. In this review, the authors explore the current literature on potential complications associated with long-term PPI use.

In: Proton Pump Inhibitors (PPIs)
Editor: Barbara Parker

ISBN: 978-1-63482-890-1

Chapter I

Proton Pump Inhibitors: Pharmacologic Principles of Use

***Abhijeet Waghray, MD*[1], *Nisheet Waghray, MD*[2], *Birju Shah, DO*[1] and *M. Michael Wolfe, MD*[*1,2]**

[1]Department of Medicine, MetroHealth Medical Center, Case Western Reserve University, OH, US

[2]Division of Gastroenterology, MetroHealth Medical Center, Case Western Reserve University, OH, US

Abstract

The secretion of large quantities of concentrated hydrochloric acid (HCl) by gastric parietal cells is a hallmark of the mammalian stomach. Since their development in the 1970s and 1980s, the mainstay of the antisecretory therapy had been histamine-2 receptor antagonists (H_2RAs). The recognition of the H^+/K^+-ATPase enzyme as the final pathway of HCl secretion led to the development of proton pump inhibitors (PPIs). PPIs are specific inhibitors of H^+/K^+-ATPase and function as prodrugs, requiring acid activation for maximum efficacy. When administered optimally, PPIs are the most potent inhibitors of gastric acid secretion,

* Address correspondence to: M. Michael Wolfe, M.D., MetroHealth Medical Center, 2500 MetroHealth Drive, G-575, Cleveland, OH 44109, Tel: 216-778-8266, Fax: 216-778-5823, E-mail: michael.wolfe2@case.edu

and their availability has expanded the therapeutic options available for acid-related disorders, including gastroduodenal ulcers, NSAID-related mucosal injury, Zollinger-Ellison syndrome, stress related ulcer syndrome in critically ill patients, gastroesophageal reflux disease, and *H. pylori* eradication therapy. As an increasing number of PPIs are being prescribed by physicians, less than optimal dosing has become an area of concern. The purpose of this chapter is to review the pathophysiology of acid secretion and inhibition, with a specific focus on the pharmacological principles surrounding the use of PPIs. Furthermore, relevant information on the management of acid-related disorders will be provided for clinicians.

Introduction

The human stomach is a specialized organ that functions to store and process food for absorption. Primarily considered a secretory organ, the stomach consists of two functional areas — the pyloric and oxyntic gland areas [1]. Parietal cells, which are the principle secretory cells in the stomach, are confined to the latter region. The secretion of large quantities of concentrated HCl (0.16 M, pH ~ 0.8) is the hallmark of the mammalian stomach and is accomplished via the nearly one billion parietal cells [2]. Secreting HCl is an energy consuming process, and 34% of parietal cell volume is thus occupied by mitochondria, representing the greatest density of mitochondria found in any cell of the human body [3]. In the basal state, the parietal cell contains numerous tubulovesicles and secretory canaliculi [4]. Stimulation following meal ingestion initiates fusion of the tubulovesicles with the surface membrane, thereby forming microvilli and dramatically increasing the cellular surface area [4]. It is this unique morphologic transformation that enables the secretion of large quantities of HCl. These changes occur rapidly over minutes and continue throughout the duration of the stimulus (e.g., meal ingestion).

H^+, K^+ adenosine triphosphatase (H^+/K^+-ATPase) catalyzes the final step of acid secretion and is responsible for the exchange of H^+ and K^+ ions by active transport across the apical membrane of the parietal cell [5]. The discovery of proton pump inhibitors (PPIs) by targeting this enzyme has transformed acid suppressive therapy. PPIs form a covalent bond to the H^+/K^+-ATPase and have supplanted H_2-receptor antagonists (H_2RAs) as the most effective inhibitors of acid secretion.

Pathways Regulating Acid Secretion (Figure 1)

Gastric acid stimulation is a highly regulated process mediated by three physiologic pathways: paracrine, neurocrine, and endocrine [2]. Enterochromaffin cells (ECLs), located adjacent to parietal cells in the oxyntic gland, produce histamine via decarboxylation of L-histadine. Histamine is the principal paracrine transmitter binding to H_2-specific receptors on the basolateral membrane of parietal cells, leading to the activation of adenylate cyclase. The resultant accumulation of intracellular 3',5'-cyclic monophosphate (cAMP) levels stimulate the generation of H^+ ions [6, 7]. Indirectly, histamine also stimulates acid secretion by binding to ECL cell H_3 receptors and suppressing the release of the inhibitory regulatory peptide somatostatin [8, 9].

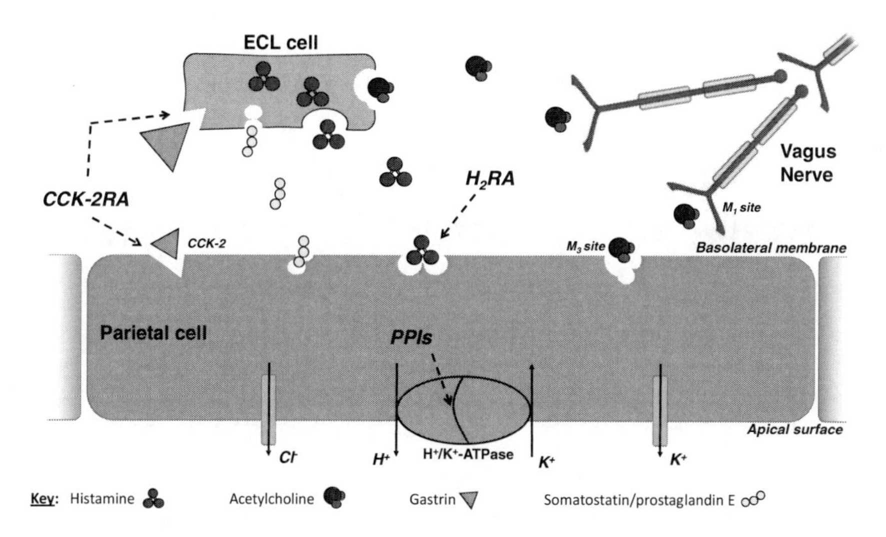

Figure 1. Factors influencing the secretion of gastric acid. Three physiologic pathways are depicted: paracrine (histamine from ECL cells and somatostatin), neurocrine (acetylcholine from vagal efferent neurons), and endocrine (gastrin secretion from G cells). Dashed arrows represent sites of acid inhibition.

Neural regulation of gastric acid secretion involves both stimulatory and inhibitory mechanisms. Acetylcholine is the principal stimulatory neurocrine transmitter released at the presynaptic terminus of the vagus nerve. Binding to the cholinergic muscarinic (M_3) receptor on the basolateral membrane of the parietal cells results in an increase in intracellular calcium leading to gastric acid secretion [10]. Neural stimulation also indirectly leads to gastric acid secretion. Acetylcholine activation of cholinergic (M_2 and M_4) receptors on somatostatin-containing D-cells inhibits somatostatin release and removes feedback inhibition of acid secretion.

Gastrin secretion by antral G-cells regulate acid secretion both directly and indirectly. Gastrin activation of the cholecystokinin-2 (CCK2) receptor on parietal cells induces an increase in intracellular calcium and directly induces acid secretion. In addition, gastrin also binds to CCK2 receptors on ECLs to stimulate the release of histamine, thereby indirectly activating parietal cells [11]. Studies in CCK2-receptor deficient mice have demonstrated that gastrin receptors on ECL cells play a major role in regulating gastric acid secretion [12, 13]. Substantial overlap exists between the three pathways to coordinate the synergistic activation and inhibition of H^+ ion generation. Inhibition of even one pathway will result in a dramatic reduction in acid secretion. Histamine appears to be the dominant mediator of acid secretion [14, 15], and intense investigation involving selective H_2 receptor inhibition culminated in the development of several H_2RAs beginning in the late 1970s, which became the mainstay of acid suppressive therapy until the development of PPIs (Figure 1).

Introduction of Proton Pump Inhibitors

By the end of the 1970s, it was clearly demonstrated that H^+/K^+-ATPase constituted the final step in acid secretion from the parietal cell [16, 17]. Subsequent studies further revealed that the chemical structure, pyridylmethylsulfinyl benzimidazole, possessed significant antisecretory properties.

By manipulating the chemical structure with substitutions to the pyridine and benzimidazole rings, drugs with higher pKa's were developed to promote their accumulation in the acidic canalicular membrane of the parietal cell, thereby increasing the rate of acid mediated conversion to the reactive species [18]. This work led to the development of a class of drugs known as PPIs, which inhibited the final pathway of acid secretion. Omeprazole was the first PPI synthesized in 1979, and its development ushered in a new era in the management of acid-related disorders by virtue of the superiority of PPIs compared to H_2RAs in their ability to suppress gastric acid secretion [19, 20].

Mechanism of Action

As mentioned above, PPIs inhibit acid secretion by targeting the final functional unit involved in acid secretion, H^+/K^+-ATPase. The parietal cell H^+/K^+-ATPase is a member of the P2-type mammalian ATPases that aid in transport of specific cations. The H^+/K^+-ATPase exists as an $(\alpha\text{-}\beta)_2$-heterodimer, consisting of an α-subunit with ten transmembrane spanning segments and a β-subunit with a single transmembrane segment. The α- and β -subunits are non-covalently linked via transmembrane sequences, allowing for the binding of transported cations (H^+) to the carboxylic amino acids in the membrane domain (17, 21-24) (Figure 2).

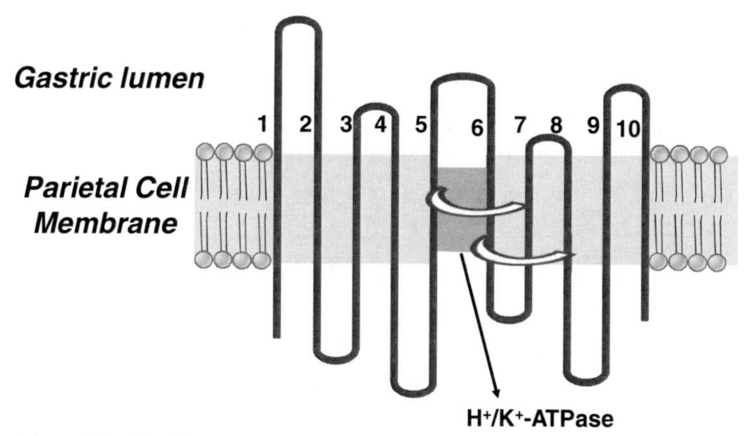

Adapted from MM Wolfe and G Sachs (2000), reference #17.

Figure 2. A model of the gastric H^+/K^+-ATPase enzyme, which is formed by an $(\alpha\text{-}\beta)_2$-heterodimer. This P2-type enzyme consists of ten transmembrane segments for the α-subunit and a single segment for the β-subunit. Segments M5 and M7, and M6 and M9 exhibit cross-linking (curved lines). H^+ ion active transport occurs between M5 and M6.

PPIs share a common structural motif, a pyridylmethylsulfinyl benzimidazole, but differ in substitutions on the pyridine or benzimidazole moieties [17]. They are all weak protonatable bases with a pKa of 5.0 for rabeprazole, pKa of 4.0 for omeprazole, esomeprazole, lansoprazole, dexlansoprazole, and pKa of 3.9 for pantoprazole. PPIs are prodrugs that traverse the parietal cell membrane and are effective only after enzymatic protonation. They suppress acid by two integral processes: 1) acid-dependent accumulation of the prodrug in the parietal cell canaliculus and 2) enzymatic conversion of the prodrug to the active thiophillic sulfenamide moiety (Figure 3).

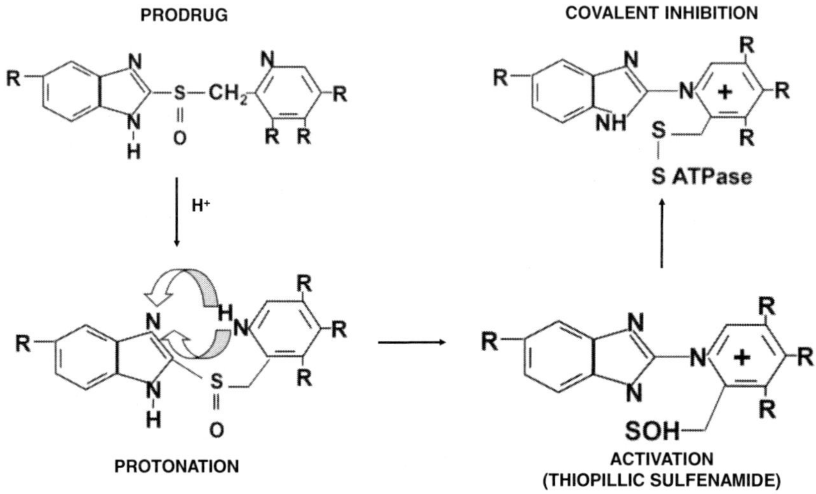

Figure 3. PPIs accumulate in the in the secretory canaliculus as prodrugs, are activated (thiophillic sulfenamide) and form a disulfide-covalent bond to the H^+/K^+-ATPase residue. Adapted from MM Wolfe and G Sachs (2000); reference #17.

Oral PPIs must be enteric coated to enable time released absorption in the duodenum. The H^+/K^+-ATPase secretes acid into the secretory canaliculus of the parietal cell, generating a regional pH of ~0.8. Once absorbed, the PPI prodrug traverses the parietal membrane as a non-protonated pyridine and accumulates in the highly acidic environment of the secretory canaliculus [17]. There, the prodrug undergoes an enzyme catalyzed conversion to the active thiophillic sulfenamide [25, 26].

As weak bases, PPIs accumulate in this highly acid environment, become activated, and covalently bind to H^+/K^+-ATPase cysteine residues. The only acidic space in the body with a pH <4 is within this organelle, so it is not surprising that PPIs accumulate in this location. The order of PPI reactivity is inversely proportional to the order of acid stability. Thus, the order of reactivity of PPIs is rabeprazole > omeprazole, esomeprazole, lansoprazole, dexlansoprazole > pantoprazole [27-29]. Studies have shown that both the R- and S-enantiomers of PPIs are converted to the active moiety, and once active, form a disulfide-covalent bond to the H^+/K^+-ATPase cysteine residue on the α-subunit, suppressing basal and stimulated acid secretion [30]. While PPIs have a circulating half-life of ~1.0-1.5 hours, this covalent bond allows for a more durable inhibition of gastric acid secretion, resulting in a biologic half-life of ~24 hours [31-33]. Interestingly, the average biologic half-life of PPIs is shorter than the half-life of the H^+/K^+-ATPase enzyme (approximately 54 hours) and varies among PPIs [34]. Evidence suggests that differences in cysteine binding sites for the active thiophillic sulfenamide may be a contributing factor. While all PPIs bind to the cysteine 813 (Figure 4), pantoprazole is unique in its ability to bind to cysteine 822, located deeper in the H^+/K^+-ATPase membrane domain [29]. *In vivo* models suggest that this additional cysteine bond could prolong the acid inhibitory properties of pantoprazole. In one study, acid secretion was restored later in those administered pantoprazole compared to omeprazole (46 hours versus 28 hours, respectively) [35, 36]. Nevertheless, the clinical relevance of this observation remains unclear.

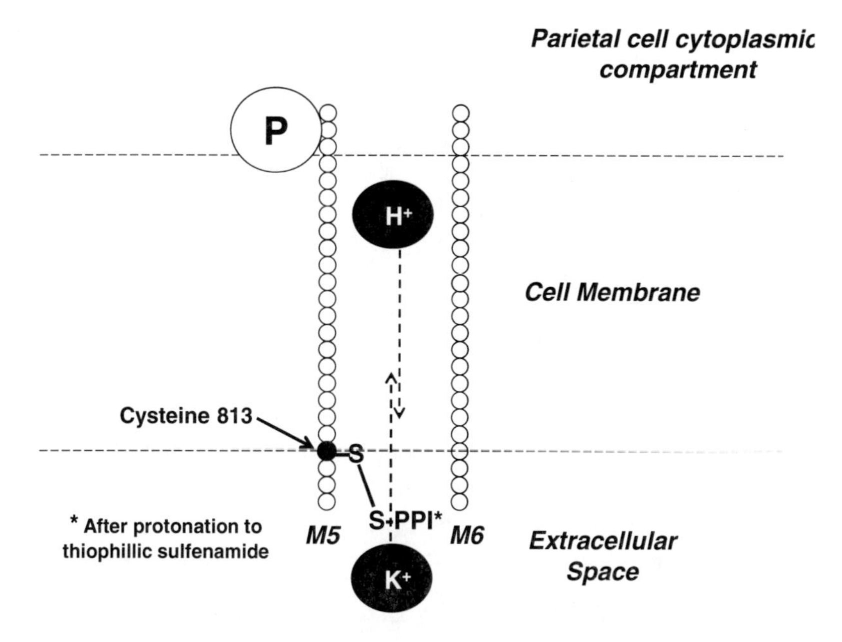

Figure 4. The enzyme structure surrounding disulfide binding of the protonated PPI (thiophillic sulfonamide) to cysteine 813 in the M5 transmembrane domain of H^+/K^+-ATPase, thereby preventing the active transport of H^+ ions in the gastric lumen in exchange for K^+ ions.

Six PPIs are currently available, including omeprazole, esomeprazole, lansoprazole, pantoprazole, rabeprazole and dexlansoprazole. All PPIs are available in oral formulation, while only esomeprazole and pantoprazole are available for intravenous administration. The bioavailability of oral PPIs varies from 60% for omeprazole to 89% for esomeprazole with repeat dosing [37, 38]. Esomeprazole is the S-enantiomer of omeprazole and dexlansoprazole is the R- enantiomer of lansoprazole, while the remaining four PPIs are an equal racemic combination of R- and S-enantiomers. All PPIs are excreted in urine with the exception of lansoprazole, which is excreted in stool. PPIs have a large therapeutic window and are safe to use in patients with renal and hepatic insufficiency without the need for dose adjustment. Except for omeprazole, which is pregnancy class C, all PPIs are labeled pregnancy class B and safe to use during pregnancy.

When administered properly, PPIs are the most effective inhibitors of acid secretion [17]. Acid suppression by PPIs relies on two integral principles: 1) the greatest amount of H^+/K^+-ATPase is present in the parietal cell after a prolonged fast and 2) acid secretion is stimulated by postprandial parietal cell activation.

Therefore, to achieve maximal efficacy, the timing of administration is critical. PPIs should be administered before the first meal of the day after a prolonged fast, followed by food ingestion to stimulate parietal cells [17]. Postprandial stimulation of parietal cells activates approximately 70-80% of all proton pumps [39]. For most patients, once daily dosing provides sufficient acid suppression, but a second dose, if deemed necessary, should be administered before the evening meal [17].

There are several clinical implications of understanding the mechanism of action of PPIs. Because PPIs have a short circulating half-life (1 to 1.5 hours), they should be taken no more than 30 minutes before ingesting a meal. Meals should include gastric acid stimulants, such as protein (e.g., dairy products) or coffee. In one study, both decaffeinated and regular coffee were shown to be potent stimulators of gastric acid secretion [40]. PPIs have a delayed onset of acid suppression, as they must accumulate in the acidic environment of the parietal cell and inhibit only activated enzymes in the parietal cells. With subsequent doses, previously inactivated enzymes are recruited, with a steady state of maximal acid inhibition of 66% by 5 days [17]. Therefore, "as needed" dosing of PPIs would not be expected to provide consistent acid inhibition or control of symptoms. Restoration of acid secretion off PPIs may take equally as long because of the biologic reversibility of the disulfide bond and enzyme turnover. Because stimulation of acid secretion is critical for the activity of PPIs, they should not be administered together with H_2RAs, misoprostol, somatostatin analogues, or other anti-secretory medications [41].

An H_2RA may be considered for individuals with nocturnal breakthrough symptoms, provided that sufficient time has elapsed from the administration of the PPI. Although no minimal time interval has been determined, in clinical practice, the H_2RA may be administered at bedtime in patients taking PPIs in the morning.

For most patients, maintenance therapy with PPIs provide adequate symptom relief. However, because of concerns regarding long-term complications and the potential for drug interactions, asymptomatic patients on maintenance therapy may consider discontinuing their medication. Concerns have been raised regarding rebound gastric acid hypersecretion following PPI cessation [42]. It has been suggested that upon the discontinuation of PPIs, antral somatostatin biosynthesis and release is suppressed, leading to an increase in the release of gastrin and a loss of physiologic feedback inhibition of acid secretion [17]. No clear criteria define prolonged therapy and there is no specific regimen for discontinuing PPI therapy. In general, patients on maintenance therapy for greater than 6 months, and who are asymptomatic for at least 3 months, may be considered for discontinuation of PPI therapy. In a randomized controlled trial of 97 patients on long term PPI therapy, no significant difference was noted in the rate of PPI resumption between those using a tapering compared to a non-tapering strategy [43].

Drug Metabolism and Drug Interactions

Proton Pump Inhibitor Metabolism

While PPIs share a common motif, differences exist in their pKa, bioavailability, route of metabolism, and excretion [44]. These differences may impact the clinical efficacy of PPIs in certain ethnic populations and may be responsible for particular drug interactions. As described earlier, rabeprazole has the highest pKa, suggesting a faster onset of action, while pantoprazole binds to both cysteine 813 and 822 and is often considered to be most specific inhibitor of acid secretion. There is little evidence, however, to suggest that these differences have any clinical relevance. Drug metabolism, in contrast, may impact the clinical efficacy of PPIs in certain ethnic populations and may be responsible for particular drug interactions. With the exception of lansoprazole, which is metabolized by cytochrome 3A4 (CYP3A4), all other PPIs are primarily metabolized by cytochrome 2C19 (CYP2C19), a member of the hepatic cytochrome P450 (CYP450) family. Genetic polymorphisms in CYP2C19 have been described in various patient populations with an impact on drug metabolism [44].

In the Asian population, two mutations in CYP2C19 have been identified, resulting in a significant delay in the metabolism of PPIs (13% of Koreans; 20% of Japanese) compared to other ethnicities (2.5% of Caucasian Americans; 2% of African Americans) [44, 45]. Conversely, 66% of Caucasians are homozygous for the wild-type gene and rapidly metabolize PPIs. Given their wide therapeutic index, no adverse effects have been reported to occur as a result of these differences in metabolism [17, 46]. However, changes in the metabolism of PPIs may impact the plasma level of drug and its clinical efficacy. In a study of Japanese patients with *H. pylori* infection treated with omeprazole, those who were homozygous for a CYP2C19 mutation ("slow metabolizers") achieved 100% eradication of the bacteria, while patients who were heterozygotes or wild type homozygotes achieved only 60% and 29% cure rates, respectively [44]. Similar results in the treatment of GERD were demonstrated in patients using lansoprazole. Therefore, CYP2C19 gene mutations can impact the plasma concentrations of PPIs and these consequences may be beneficial or detrimental in terms of treatment and drug interactions.

Drug Interactions

Clopidogrel is a prodrug requiring activation via the CYP450 enzyme pathway. In particular, cytochrome P450 1A2, 2B6, 2C9, 2C19, and 3A4 are involved in the activation of clopidogrel, with the latter two iso-enzymes assuming the greatest clinical significance [47, 48]. Given their shared metabolic pathway, concerns of drug interactions with the concomitant use PPIs and clopidogrel have been raised. While the exact mechanism remains unclear, competitive inhibition of CYP2C19, genetic polymorphisms in CYP450, and elevated plasma levels of asymmetric dimethylarginine have been proposed as potential etiologies.

Research over the years has provided conflicting results. Initial *in vitro* and *in vivo* data demonstrated that PPIs decreased clopidogrel inhibitory effects on platelets [49, 50]. In 2009, a retrospective analysis of 8,205 patients demonstrated that clopidogrel-PPI therapy after a coronary event resulted in a 9% increased risk of re-hospitalization or death from acute coronary syndrome (ACS) when compared to clopidogrel alone [51].

A subsequent case controlled study and an evaluation of patients in a French registry yielded conflicting results, although, these studies were limited by several confounding factors [52, 53].

Clopidogrel and the Optimization of Gastrointestinal Events Trial (COGENT) is the only randomized controlled trial to compare PPI (omeprazole) to placebo in patients taking clopidogrel/aspirin (DAPT). The data published in late 2010 demonstrated no significant difference in the risk of cardiovascular events with combination therapy [54]. A consensus recommendation by a multi-society task force of gastroenterologists and cardiologists advised that patients on clopidogrel should consult with their physicians regarding the risk and benefit of concomitant PPI therapy [55].

PPIs also demonstrate interactions with several medications and herbal supplements. Serum concentrations of diazepam, phenytoin, warfarin, theophylline, delayed-release risedronate (bisphosphonate), pimozide (antipsychotic medication) and antiretroviral agents (e.g., delavirdine) are reduced in the setting of PPI use, and caution must be exercised during concomitant administration [56-59]. In contrast, St John's Wort, ginkgo biloba, yin zhihuang, and artemisinin reduce serum concentrations of PPI by as much as 20-50% and should thus be avoided [60, 61]. Furthermore, the use of newer agents for treating hepatitis C may necessitate dose adjustment in PPI therapy. Medications dependent on low pH for solubility (e.g., ledipasvir/sofosbuvir) should not be taken with more than 20 mg of omeprazole or its equivalent per day, while patients receiving a co-packaged formulation of ombitasvir/partaprevir/ritonavir/dasabuvir may require an increase in omeprazole dosing [62, 63].

Treatment Involving Proton Pump Inhibitors

PPIs inhibit acid secretion and promote mucosal healing superior to any other class of acid suppressive therapy. They have a large therapeutic window and are safe in a variety of clinical situations. Their availability has expanded the therapeutic options available for acid-related disorders, including gastroduodenal ulcers, NSAID-related mucosal injury, gastroesophageal reflux disease (GERD), Zollinger-Ellison syndrome, and stress related ulcer syndrome in critically ill patients (Table 1).

Table 1. Dosing of PPIs. For treating and primary/secondary prevention of gastroduodenal ulcers$^{\alpha}$ (4 and 8 weeks duration of treatment for active gastric and duodenal ulcers, respectively), and treating gastroesophageal reflux disease$^{\beta}$

Proton Pump Inhibitor	Adult Dose (oral): Administered once daily before breakfast, with second dose before evening meal (if needed)
Dexlansoprazole	30-60 mg once daily$^{\alpha}$ 30 mg once daily or 30 mg twice daily$^{\beta}$
Esomeprazole	20-40 mg once daily$^{\alpha}$ 20 or 40 mg once daily$^{\beta}$
Lansoprazole	15-30 mg once daily$^{\alpha}$ 30 mg once daily or 30 mg twice daily$^{\beta}$
Omeprazole	20-40 mg once daily$^{\alpha}$ 20-40 mg once daily or 20 mg twice daily$^{\beta}$
Pantoprazole	20-40 mg once daily$^{\alpha}$ 40 mg once daily or 40 mg twice daily$^{\beta}$
Rabeprazole	20 mg once daily$^{\alpha}$ 20 mg once daily or 20 mg twice daily$^{\beta}$

Peptic Ulcer Disease

Regardless of the etiology of peptic ulcer disease (PUD) (e.g., *H. pylori* infection, NSAID use, hypersecretory states), acid suppression remains the cornerstone of therapy. Comparative trials have shown that PPIs are more effective and heal gastroduodenal ulcers more rapidly than H_2RAs; however, there is no difference in healing ulcers among the PPIs. In a meta-analysis of 30 trials evaluating the healing of duodenal ulcers, 20 mg of omeprazole every morning for four weeks was superior to both 300 mg of ranitidine and 800 mg of cimetidine, both administered at bedtime [64]. Subsequently, a meta-analysis of five trials showed that, compared to ranitidine and famotidine, lansoprazole healed significantly more ulcers, with healing rates of 60% and 85% for lansoprazole at two and four weeks, respectively, compared to 40% and 75% for H_2RAs [65]. These results were further confirmed with studies involving both rabeprazole and pantoprazole [66, 67]. Similar findings were observed in the treatment of gastric ulcers, with PPIs demonstrating greater efficacy compared to H_2RAs [64]. The optimal duration of therapy for duodenal and gastric ulcers should be four and eight weeks, respectively, for both PPIs and H_2RAs.

Nonsteroidal Anti-Inflammatory Drug (NSAID) Associated Mucosal Injury

NSAIDs are widely prescribed for analgesia and musculoskeletal disorders, but their use is commonly associated with gastrointestinal mucosal injury (e.g., erosions, ulcers, and strictures). Even short term therapy of less than 3 months can result in significant gastroduodenal injury [68]. Nonselective NSAIDs inhibit both cyclooxygenase (COX) -1 and -2, thereby reducing endogenous prostaglandin synthesis and rendering the upper gastrointestinal tract vulnerable to acid-related mucosal injury [69]. Risk factors for NSAID-related mucosal injury include advanced age, high dose of NSAIDs, history of gastroduodenal ulcer/gastrointestinal bleeding, concurrent steroid/anticoagulant use, and significant co-morbid conditions. The optimal management of NSAID-induced mucosal injury is to discontinue NSAIDs or initiate concomitant therapy with a PPI.

Misoprostol is a prostaglandin analog approved for the prevention of NSAID-induced gastroduodenal ulcers. PPIs are equal to or more effective than misoprostol in healing NSAID-induced ulcers. In a double-blind, randomized study of 935 patients taking NSAIDs, those who received omeprazole experienced better rates of ulcer healing and a 13% lower rate of relapse compared to misoprostol [70]. The use of misoprostol is further limited by its frequency of dosing, contraindication in pregnancy, and diarrhea, the latter observed in a dose-dependent manner in 20-30% of those taking the drug.

Several multicenter trials have demonstrated the superiority of PPIs to H_2RAs in healing NSAID-related erosions/ulcers. In one study, patients with gastroduodenal ulcers who continued NSAID therapy were evaluated for resolution of ulcers/erosions on omeprazole or ranitidine. Ulcer healing rates at 8 weeks were 79% and 80% for patients receiving omeprazole 40 mg or 20 mg, respectively compared to 63% in those administered ranitidine 150 mg twice daily (Figure 5). At the end of 6 months, remission rates were 13% higher in the omeprazole group, with no increased risk of adverse events [71]. These results were further corroborated in a study comparing the use of lansoprazole 15 mg and 30 mg to ranitidine 150 mg twice daily to assess healing of gastric ulcers in patients continuing NSAID use. Patients treated with lansoprazole 15 mg and 30 mg demonstrated higher rates of healing compared to patients treated with ranitidine (69%, 73%, and 53%, respectively) [72].

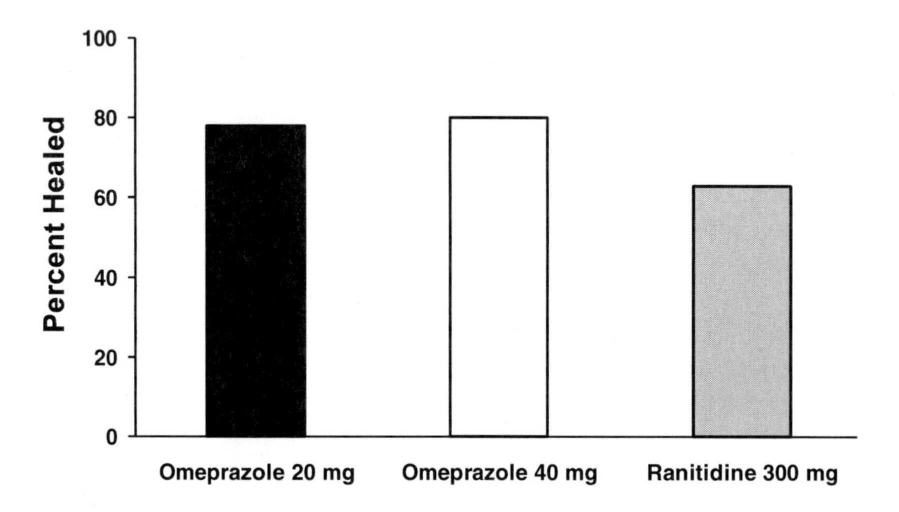

Figure 5. Comparison of patients with NSAID-associated gastroduodenal ulcers treated with omeprazole versus ranitidine while continuing NSAID therapy, demonstrating the superiority of PPIs compared to H2RAs. Adapted from Yeomans, *et al.* (1998), reference #71.

PPIs have further proven to be effective in the primary and secondary prevention of NSAID-related mucosal injury and ulcer-related complications. Two large randomized multicenter studies (VENUS and PLUTO) evaluated the utility of using esomeprazole for ulcer prophylaxis in patients taking NSAIDs. Pooled data revealed ulcer rates of 16.5% in patients receiving placebo, compared to 0.9% and 4.1% in patients receiving esomeprazole 20 mg and 40 mg, respectively [73]. Secondary prevention was further evaluated in a study involving arthritis patients with a history of a healed gastroduodenal ulcer and continued NSAID therapy. A total of 432 patients were randomly stratified to two treatment groups, omeprazole 20 mg daily and ranitidine 150 mg twice daily. Analysis at 6 months revealed recurrent gastric and duodenal ulcers occurring in 5.2% and 0.5% of patients who received omeprazole, respectively compared to 16.3% and 4.2% of patients receiving ranitidine [71]. The OBERON trial also demonstrated PPI superiority in the prevention and treatment of dyspeptic symptoms related to the use of low-dose aspirin [74].

Overall, these studies demonstrate the superiority of PPIs compared to misoprostol and H_2RAs in the management of NSAID-related mucosal injury and ulcer-related complications.

Bleeding Ulcers – A Complication of Peptic Ulcer Disease

Numerous studies have evaluated the role of acid suppressive therapy in upper gastrointestinal bleeding due to gastroduodenal ulcers. In a meta-analysis of 21 randomized controlled trials, medical management for bleeding ulcers with or without endoscopic therapy revealed a significant reduction in episodes of re-bleeding (OR 0.46, 95% CI 0.33 – 0.64) and need for surgical intervention (OR 0.59, 95% CI 0.46 – 0.76) [75]. In general, the use of H_2RAs in ulcer-related hemorrhage have yielded conflicting results, particularly in the management of bleeding duodenal ulcers [76-80]. In a systemic review of 30 trials, H_2RAs demonstrated a minor reduction in the risk of re-bleeding from gastric ulcers and no benefit in bleeding duodenal ulcers when compared to placebo [79]. Trials comparing PPIs and H_2RAs have shown that PPIs are more effective in the treatment of bleeding gastroduodenal ulcers. In a meta-analysis of 9 trials with H_2RAs or placebo as the control, PPI therapy significantly reduced the rates of re-bleeding (OR 0.50, 95% CI 0.33 – 0.77) and need for surgical intervention (OR 0.47, 95% CI 0.29 – 0.77) (81). The rationale for these findings lies in the ability of PPIs to maintain gastric pH > 6.0, decreasing the proteolytic activity of pepsin and stabilizing clot formation [82].

Continuous versus Intermittent PPI Therapy

Current guidelines by the American College of Gastroenterology, which were established in 2012, recommend continuous high-dose PPI infusion (e.g., esomeprazole or pantoprazole: 80 mg bolus followed by 8 mg per hour infusion) in patients with ulcers containing high-risk stigmata of recent hemorrhage following endoscopic therapy. Several recent studies comparing continuous infusion to intermittent dosing have failed to show any difference in clinical endpoints. In a meta-analysis of seven randomized controlled trials, recurrent bleeding rates, the need for operative management, and mortality were compared in 1,157 patients who received endoscopic therapy and high-dose PPI (esomeprazole or pantoprazole given as 80 mg bolus followed by 8 mg/hour) or lower doses of PPIs (esomeprazole or pantoprazole with oral dosing ranging from 80 to 160 mg per day or IV ranging from 20 mg to 80 mg per day). No difference was noted in outcomes between the two groups [83]. A more recent meta-analysis of 1,716 patients detected no difference between high-dose PPI (defined as greater than 600 mg of IV PPI therapy in 72 hours) and low-dose PPI therapy in terms of hospital length of stay or blood transfusion requirements [84]. A meta-analysis published in 2014 compared continuous PPI infusion to intermittent dosing for high-risk bleeding ulcers. Intermittent dosing was comparable to continuous infusion in the primary (recurrent bleeding at 7 days) and secondary (recurrent bleeding between 3 and 30 days) outcomes, need for transfusion, and mortality [85].

Oral versus Intravenous PPI Therapy

Intravenous PPI therapy is considered to be the standard of care for gastroduodenal ulcers with high-risk stigmata of recent hemorrhage, however several studies have evaluated the role of oral therapy as an alternative approach. Oral therapy is an intriguing alternative because of its lower cost compared to continuous or intermittent IV PPI therapy. In one study, after endoscopic therapy for bleeding ulcers, patients were randomized to receive either continuous IV lansoprazole infusion (90 mg bolus followed by 9 mg/hour) or oral lansoprazole (120 mg initial dose followed by 30 mg every 3 hours). While both groups achieved a gastric pH >6 in more than 60% of patients, patients receiving IV PPI attained this goal 1 hour sooner than those on oral therapy. The clinical significance of a more rapid increase in gastric pH remains unclear [86]. A meta-analysis of 615 patients with bleeding peptic ulcer disease demonstrated no significant difference in recurrent bleeding, blood transfusion volume, mortality, or need for operative management between oral and IV PPI therapy [87]. Most recently, a single center, double-blind trial involving 244 patients compared IV bolus plus infusion of esomeprazole to high-dose oral esomeprazole therapy in patients with peptic ulcers with high-risk stigmata of recent hemorrhage. There was no difference in blood transfusion, repeat endoscopic therapy, or hospital stay between the two groups [88].

PPIs clearly demonstrate greater efficacy in ulcer healing and are the standard of care in the management of peptic ulcer disease with high-risk stigmata of recent hemorrhage. Studies suggest that intermittent and oral PPI therapy may be alternatives to continuous IV PPI infusion following endoscopic treatment of high-risk bleeding ulcers. Certainly, provided patients are able to ingest medication, oral therapy would reduce overall resource utilization and cost of care, but further large randomized controlled trials are required to assess the efficacy of oral PPI therapy in the management of high-risk bleeding peptic ulcers.

Stress-Related Ulcers

Stress-related gastroduodenal lesions (e.g., superficial ulcers, erosions, or subepithelial hemorrhages) can develop rapidly during periods of physiologic stress. In critically ill patients there is a marked imbalance between mucosal protection and gastric acid secretion [89, 90]. Studies have reported endoscopic evidence of mucosal damage in >75% of critically ill patients, and the severity of disease correlates with the risk of developing gastroduodenal lesions [89]. The incidence of gastrointestinal bleeding secondary to stress ulcer syndrome (SUS) ranges from 1.5% to 8.5% of all ICU patients to as high as 15% in those not prescribed stress ulcer prophylaxis [91-94]. Independent risk factors for SUS include mechanical ventilation (at least 48 hours) and coagulopathy (International Normalized Ratio (INR) or prothrombin time > 1.5, platelet count <50,000, or a partial thromboplastin time (PTT) more than twice the control value) [91]. The risk of significant gastrointestinal bleeding with >1 risk factor is 3.7% compared to 0.1% in those with no risk factors [91].

Although no mortality benefit has been identified, prophylactic therapy with H_2RAs or PPIs reduces the incidence of bleeding related to stress-related mucosal injury. Current guidelines state that stress ulcer prophylaxis should be provided to critically ill patients with risk factors for bleeding [90, 95]. Studies comparing the intravenous administration of H_2RAs and PPIs suggest that while both agents raise intragastric pH to >4, PPIs are more effective at maintaining a pH >6 without the issues related to tolerance seen with IV H_2RA use. Increasing and maintaining intragastric pH levels >6 reduces the conversion of pepsinogen to pepsin, enables the clotting cascade, and allows for clot stabilization; all of these factors are critical for hemostasis and for achieving a reduction in the risk for SUS. Data from a meta-analysis of 13 trials including nearly 1,600 critically ill patients demonstrated that PPIs significantly decreased the risk of bleeding compared to H_2RAs (OR 0.39; 95% CI: 0.19–0.77) [96]. PPIs are not only superior in the prevention of SUS, but are also likely cost effective. A recent cost-effective analysis compared the use of PPIs (IV bolus or oral omeprazole 40 mg) to H_2RAs (IV famotidine 40 mg twice daily) in patients admitted to ICUs. Comparing preventative costs and cost per complication (bleeding and ventilator associated pneumonia), PPIs appear to represent the safer and more cost effective therapeutic option [97].

Gastroesophageal Reflux Disease (GERD)

Approximately 44% of the adult US population experiences heartburn symptoms at least once per month, resulting in a significant cost to the healthcare system [98-100]. Numerous studies have evaluated the efficacy of antisecretory therapy in healing mucosal injury and providing symptom relief. At standard dosing (omeprazole 20 mg, lansoprazole 30 mg, rabeprazole 20 mg, pantoprazole 40 mg), PPIs administered before breakfast heal esophagitis and provide symptomatic relief in 85-90% of patients. Non-response to standard therapy can be seen in patients with more severe esophagitis, suboptimal PPI administration, and individual pharmacogenomic differences resulting in slow or rapid metabolism of PPIs [101, 102].

Comparative trials suggest that PPIs are more effective than H_2RAs in the management of GERD and its related complications [103, 104]. In one trial, patients with moderate to severe erosive GERD were treated with lansoprazole 30 mg daily or ranitidine 300 mg twice daily, with healing rates at 8 weeks of 91% and 66%, respectively [105]. A large meta-analysis including 43 studies demonstrated that PPIs were nearly twice as fast as H_2RAs in achieving complete resolution of esophagitis and heartburn symptoms (103) (Figure 6). Similarly, a recent Cochrane review of patients with non-erosive GERD revealed that PPIs (RR 0.37; 95% CI 0.32-0.44) were more effective at relieving heartburn symptoms compared to H_2RAs (RR 0.77; 95% CI 0.60-0.99) [106]. Patients with esophagitis unresponsive to high-dose H_2RAs (cimetidine 800 mg four times daily, ranitidine 300 mg three times daily) also showed improved healing and symptom control when administered omeprazole 40 mg daily [107].

Comparative trials of the available PPIs demonstrate similar efficacy in the relief of GERD symptoms. A meta-analysis of randomized controlled trials evaluating omeprazole, lansoprazole, pantoprazole, and rabeprazole in the treatment of erosive esophagitis found no significant difference in symptom relief or in the rate of mucosal healing [108].

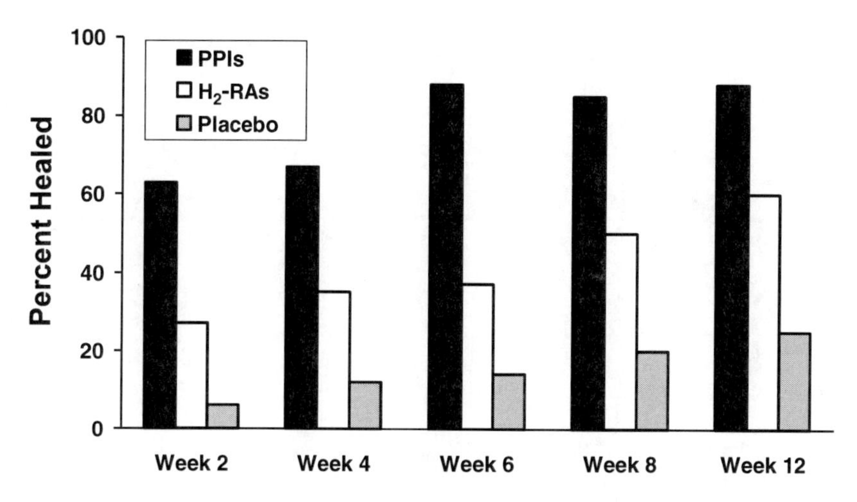

Figure 6. Meta-analysis of healing of erosive esophagitis with PPIs compared to H2RAs and placebo. The data demonstrate superior and more rapid healing with PPIs, with 2-week PPI healing rates greater than 12-week healing with H2RAs. Adapted from Chiba, et al. (1997), reference #103.

Esomeprazole, the active enantiomer of omeprazole, has been the subject of several trials comparing it to other PPIs in symptom resolution and healing of esophagitis. A multicenter randomized double blind trial of 1,960 patients with endoscopically confirmed reflux esophagitis demonstrated that 8 weeks of therapy with esomeprazole at 20 mg and 40 mg daily was more effective than omeprazole 20 mg daily for symptom relief and resolution of esophagitis (94.1% vs 89.9% vs 86.9%, $p < 0.05$, respectively) [109]. Another randomized controlled trial of 2,425 patients showed significantly better rates of healing esophagitis and symptom control in patients administered esomeprazole (93.7%) compared to omeprazole (84.2%) [110]. Because esomeprazole contains only the active enantiomer in omeprazole, these results are not surprising. Studies comparing esomeprazole to other PPIs have demonstrated no significant superiority among the PPIs when administered in equivalent doses. In a study of 5,241 patients, esomeprazole demonstrated similar absolute healing rates when compared to lansoprazole [111].

Data does not support substituting an alternative PPI in non-responders. GERD patients refractory to daily lansoprazole were randomized to receive esomeprazole therapy daily or to increase their current medical regimen to twice daily. Increasing the dose to twice daily was as effective as changing PPI therapy [112]. PPIs are also more effective in treating GERD-related complications, such as esophageal strictures. Patients treated with PPIs for peptic esophageal strictures showed a significant improvement in dysphagia related symptoms and a decreased need for repeat dilation when compared to those receiving H_2RAs [113, 114].

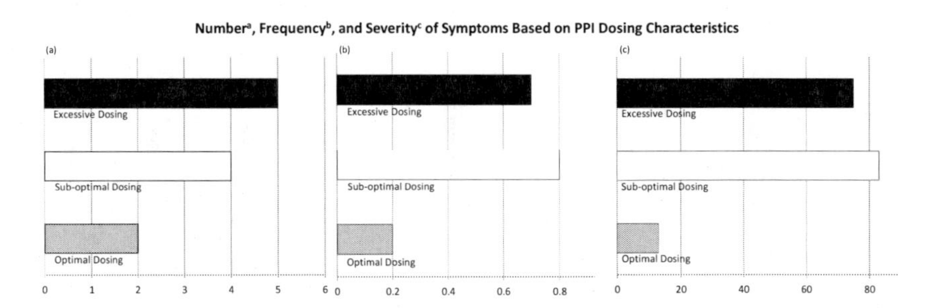

Figure 7. Improvement in GERD symptoms (number, frequency and severity) with PPIs dosed appropriately. Data presented as Gastroesophageal Reflux Disease Symptom Assessment Scale (GSAS) scores. Adapted from Sheikh, *et al.* (2014), reference #101.

Patients with GERD and its related complications generally require long-term therapy. The majority of patients, especially those with moderate to severe esophagitis, relapse upon discontinuation of therapy [115]. Maintenance therapy with PPIs is superior to all other acid suppression strategies in achieving remission and generally require the same dose that was required to achieve mucosal healing [116]. As previously discussed, PPIs should be ideally administered before the first meal to enable rapid recruitment of H^+/K^+-ATPase molecules following an overnight fast. Suboptimal dosing in clinical practice is common, with one survey reporting that only one-third of primary care physicians dosed patients optimally [117]. Similar concerns apply to over-the-counter (OTC) users of PPIs. A recent study of 1,959 patients demonstrated that OTC users and patients prescribed by primary care physicians were less likely to be optimally dosed, with worse symptoms and higher frequency and severity scores reported when compared to patients receiving prescriptions from gastroenterologists [101]. These studies demonstrate that GERD symptoms improve with appropriate PPI administration (Figure 7). In patients who respond suboptimally to PPIs, a careful history regarding the time of administration should be obtained.

Zollinger Ellison Syndrome

Zollinger Ellison syndrome (ZES) is a disorder characterized by a slow-growing tumor (gastrinoma) and comprises a clinical triad that includes hypergastrinemia, gastric acid hypersecretion, and a gastrin-secreting neuroendocrine tumor typically located in the pancreas or duodenal wall. In addition to stimulating acid secretion, gastrin also possesses trophic properties in the GI tract, including an effect on parietal cells and histamine-secreting ECLs. The trophic effect of hypergastrinemia results in an increased parietal cell mass, and consequently, patients with ZES experience a marked increase in basal and maximal acid output, with total acid output often exceeding 10 L/day [118], causing gastroduodenal ulceration in >90% of patients. Before the advent of potent acid suppressive therapies, the primary treatment for ZES was total gastrectomy, a procedure associated with significant morbidity [119]. In the late 1970s, H_2RAs provided a therapeutic alternative to surgery. However, for these medications to be effective, they needed to be administered at doses associated with significant adverse effects (mean dose: cimetidine, 3.6 g/day; ranitidine, 1.2 g/day; famotidine, 0.25 g/day) [120]. Furthermore, the short duration of action and risk of tachyphylaxis limited the role of H_2RAs. The discovery of PPIs provided a safe and reliable measure of acid suppression with minimal adverse effects, thereby transforming the management of ZES [121].

ZES treatment should commence with twice the usual PPI dose used to treat NSAID- or *H. pylori*-associated ulcers (e.g., omeprazole 40 mg) [122]. Once or twice daily dosing is sufficient, with a median effective dose of 80 mg of omeprazole [121, 123]. Lansoprazole is the most extensively studied PPI treatment for ZES with the longest reported follow-up period [124, 125]. Abdominal pain is the most common reported symptom of ZES and may be related to peptic ulcer and/or GERD. Clinical resolution of symptoms, such as relief of epigastric pain, does not reliably correlate with mucosal injury in ZES [126]. Optimal acid suppression (not achlorhydria) with a basal acid output (BAO) of 1-10 mmol/h is the only parameter proven reliable in promoting and maintaining mucosal healing [126]. The BAO should be ideally measured after PPI therapy achieves a steady state and evaluated about one hour prior to the subsequent dose of PPI. For a BAO exceeding 10 mmol/h, the PPI dose should be increased incrementally. If the PPI dose exceeds 60 mg of omeprazole or its equivalent, the dosing should be divided equally before breakfast and dinner. Dose adjustments should made every 6-12 months based on BAO, with less than 10% of patients requiring PPI dose increase [126].

Safety/Adverse Effects of Proton Pump Inhibitors

Long-term maintenance therapy and an overall increase in PPI use has prompted concerns over their safety profile. Common side effects range from 1% to 10% and include headache, nausea, vomiting, abdominal pain, diarrhea, and constipation, while serious adverse reactions are rare. The main concerns over maintenance PPI use include prolonged hypochlorhydria, hypergastrinemia, and the possibility of chronic atrophic gastritis. Specifically, safety concerns exist regarding hypochlorhydria and the potential for infections such as *Clostridium difficile*-associated diarrhea (CDAD), community acquired pneumonia (CAP), and malabsorption. Recent studies have further raised issues regarding infectious complications, electrolyte disturbances, and metabolic bone disease related to chronic PPI use.

Infections

Gastric acidity protects against bacterial overgrowth and enteric infections, and it would therefore seem logical that the inhibition of gastric acid would increase the potential for enteric infections [71, 72]. In a systematic review of 2,948 patients, those receiving PPI therapy were at an increased risk of enteric infection compared to patients receiving either H_2RAs or no anti-secretory therapy [127]. However, given the heterogeneity among the studies, any link of causation remains unclear.

C. difficile is a spore-forming anaerobic organism, and its acid resistant spores are thought to be the source of transmission. While several studies have implicated PPI use as a risk factor for CDAD, the pathophysiology of acid suppression and the potential for *C. difficile* infection remains uncertain. In a recent meta-analysis involving approximately 300,000 patients and 42 observational studies, PPI use was associated with an increased risk of CDAD (OR 1.74; 95% CI 1.47-2.85) compared to those not on therapy [128]. Pooled analysis also demonstrated an increased risk of recurrent *C. difficile* infection (OR 2.51; 95% CI 1.16-5.44). Despite these results, significant heterogeneity existed between the study groups and could not be explained on sub-group analysis. Two additional studies also demonstrated an increased risk of *C. difficile* infection with PPI use [127, 129]. Given these findings, the United States Food and Drug Administration (FDA) issued an alert to providers to consider the diagnosis of *C. difficile* infection in cases of concurrent PPI use and persistent diarrhea, as well as recommendations to consider the lowest dose and duration of PPI therapy appropriate for treatment [130]. However, as stated above, *C. difficile* is an anaerobic organism that sporulates, and these acid-resistant spores are presumed to be the major vector of disease transmission. The role of potent acid suppression and the pathophysiologic mechanisms involved in increasing the risk of *C. difficile* by PPI use thus remain unclear.

Observational studies have further proposed that hypochlorhydria secondary to PPI use may permit organisms to colonize the upper gastrointestinal tract, resulting in an increased risk of both CAP and healthcare-associated pneumonia (HCAP) [131-133]. In a 2011 meta-analysis of 31 studies, acid suppression was associated with an increased risk of developing pneumonia in those receiving PPIs (OR 1.27; 95% CI 1.11-1.46) and H_2RAs (OR 1.22; 95% CI 1.09-1.36) [134]. A large case controlled study further demonstrated an increased risk of CAP within 30 days of initiating PPI with the greatest risk within the first 48 hours of initiation [133]. While these studies suggest an association with PPI use and CAP, other health-related issues may be prevalent in those using PPIs that may predispose them to pneumonia [135, 136]. A more recent study has challenged these findings, and by using a more restricted cohort of patients attempted to minimize confounding factors. Patients were prescribed PPIs (n = 96,870) or H_2RAs (n = 47,344) for NSAID prophylaxis, and no increased risk of hospitalization for pneumonia was detected (OR 1.05; 95% CI 0.89-1.25) [137].

Malabsorption

Acid is integral to the absorption of iron, calcium, magnesium, and vitamin B12. While the impact on iron and vitamin B12 may be negligible, hypomagnesemia and metabolic osteodystrophy may be of greater clinical significance [138]. Theoretically, hypochlorhydria may decrease calcium absorption and may inhibit osteoclast activity [139, 140]. Prior studies evaluating both the dose and duration of PPI use with fracture risk revealed conflicting results [141-144]. In a meta-analysis of 11 studies involving 1,084,560 patients with 62,210 patients receiving PPIs, there was a 1.30-, 1.56-, and 1.16-fold increased risk of hip, spine and any site fractures, respectively [145]. A subsequent analysis by the Nurse's Health Study yielded similar results. In this analysis of 79,899 postmenopausal women, the absolute risk of hip fracture in patients using PPI was 2.02 events per 1,000 person years compared to 1.51 events per 1,000 patient years in those not using PPIs. Interestingly, current and former smokers treated with PPIs had a 51% increase in hip fractures (HR 1.51; 95% CI 1.0-1.91) compared to no significant difference among non-smokers using PPIs [146]. Thus, the possibility of other biological factors unrelated to PPI use impacting the risk of fractures cannot be excluded. Nevertheless, the FDA also advises clinicians to be mindful of the risk of fractures when prescribing long-term PPIs at high doses [147]. Overall, clinicians should be aware that chronic PPI therapy may result in decreased absorption of calcium, vitamin B12, iron, and magnesium. Periodic monitoring with supplementation in select patients is recommended [148]. As in all clinical situations, the risk/benefit analysis must be considered when using PPIs (or any medication). Initial concerns for PPI related hypergastrinemia and gastric carcinoid tumors demonstrated in rodent models have not been confirmed in human studies (149, 150). Furthermore, the theoretical risk of hypergastrinemia and colon cancer has not been established, with one study demonstrating no increased risk of colon cancer with PPI use [151]. Retrospective analysis and case reports have also suggested an association between PPI use and spontaneous bacterial peritonitis, acute interstitial nephritis, and gastric polyps, although data remains scarce and the mechanism not fully understood [152-157].

Conclusion/Review

PPIs inhibit the final pathway of acid secretion, H^+/K^+-ATPase, and are the most potent acid suppressive agents available. Their availability via prescription and OTC has expanded the therapeutic options for treating acid-related disorders. As an increasing number of PPIs are being utilized, concerns regarding suboptimal dosing, drug interaction, and long-term safety have been raised. In general, PPIs appear to be safe and the most cost effective option for a variety of acid-related disorders.

References

[1] Nompleggi DJ, Wolfe M.M. Peptic ulcer disease - pathogenesis and treatment. In: In Zakim D, Dannenberg, A.J., editor. Peptic Ulcer

Disease and Other Acid-Related Disorders. Armonk, NY: Academic Research Associates.33-66.

[2] Wolfe MM, Soll AH. The physiology of gastric acid secretion. *N. Engl. J. Med.* 1988;319(26):1707-15.

[3] Boeckxstaens GE, Smout A. Systematic review: role of acid, weakly acidic and weakly alkaline reflux in gastro-oesophageal reflux disease. *Aliment Pharmacol. Ther.* 2010;32(3):334-43.

[4] Sawaguchi A, McDonald KL, Forte JG. High-pressure freezing of isolated gastric glands provides new insight into the fine structure and subcellular localization of H+/K+-ATPase in gastric parietal cells. *J. Histochem. Cytochem.* 2004;52(1):77-86.

[5] Sachs G, Chang HH, Rabon E, Schackman R, Lewin M, Saccomani G. A nonelectrogenic H+ pump in plasma membranes of hog stomach. *J. Biol. Chem.* 1976;251(23):7690-8.

[6] Soll AH, Wollin A. Histamine and cyclic AMP in isolated canine parietal cells. *Am. J. Physiol.* 1979;237(5):E444-50.

[7] Soll AH. Potentiating interactions of gastric stimulants on [14 C] aminopyrine accumulation by isolated canine parietal cells. *Gastroenterology.* 1982;83(1 Pt 2):216-23.

[8] Vuyyuru L, Harrington L, Arimura A, Schubert ML. Reciprocal inhibitory paracrine pathways link histamine and somatostatin secretion in the fundus of the stomach. *Am. J. Physiol.* 1997;273(1 Pt 1):G106-11.

[9] Vuyyuru L, Schubert ML, Harrington L, Arimura A, Makhlouf GM. Dual inhibitory pathways link antral somatostatin and histamine secretion in human, dog, and rat stomach. *Gastroenterology.* 1995;109(5):1566-74.

[10] Chan FK, Wong VW, Suen BY, Wu JC, Ching JY, Hung LC, et al. Combination of a cyclo-oxygenase-2 inhibitor and a proton-pump inhibitor for prevention of recurrent ulcer bleeding in patients at very high risk: a double-blind, randomised trial. *Lancet.* 2007;369(9573):1621-6.

[11] Kopin AS, Lee YM, McBride EW, Miller LJ, Lu M, Lin HY, et al. Expression cloning and characterization of the canine parietal cell gastrin receptor. *Proc. Natl. Acad. Sci. U S A.* 1992;89(8):3605-9.

[12] Langhans N, Rindi G, Chiu M, Rehfeld JF, Ardman B, Beinborn M, et al. Abnormal gastric histology and decreased acid production in cholecystokinin-B/gastrin receptor-deficient mice. *Gastroenterology.* 1997;112(1):280-6.

[13] Samuelson LC, Hinkle KL. Insights into the regulation of gastric acid secretion through analysis of genetically engineered mice. *Annu. Rev. Physiol.* 2003;65:383-400.
[14] Prinz C, Scott DR, Hurwitz D, Helander HF, Sachs G. Gastrin effects on isolated rat enterochromaffin-like cells in primary culture. *Am. J. Physiol.* 1994;267(4 Pt 1):G663-75.
[15] Waldum HL, Sandvik AK, Syversen U, Brenna E. The enterochromaffin-like (ECL) cell. Physiological and pathophysiological role. *Acta Oncol.* 1993;32(2):141-7.
[16] Forte JG, Lee HC. Gastric adenosine triphosphatases: a review of their possible role in HCl secretion. *Gastroenterology.* 1977;73(4 Pt 2):921-6.
[17] Wolfe MM, Sachs G. Acid suppression: optimizing therapy for gastroduodenal ulcer healing, gastroesophageal reflux disease, and stress-related erosive syndrome. *Gastroenterology.* 2000;118(2 Suppl 1):S9-31.
[18] Snaeder W. The Development of a New Proton-Pump Inhibitor: The Case History of Pantoprazole in: Drug prototypes and their exploitation. 2006;Senn-Bilfinger, Jörg; Sturm, Ernst 414-5.
[19] Munson K, Garcia R, Sachs G. Inhibitor and ion binding sites on the gastric H,K-ATPase. *Biochemistry.* 2005;44(14):5267-84.
[20] Fellenius E, Berglindh T, Sachs G, Olbe L, Elander B, Sjostrand SE, et al. Substituted benzimidazoles inhibit gastric acid secretion by blocking (H+ + K+)ATPase. *Nature.* 1981;290(5802):159-61.
[21] Vagin O, Denevich S, Munson K, Sachs G. SCH28080, a K+-competitive inhibitor of the gastric H,K-ATPase, binds near the M5-6 luminal loop, preventing K+ access to the ion binding domain. *Biochemistry.* 2002;41(42):12755-62.
[22] Vagin O, Munson K, Lambrecht N, Karlish SJ, Sachs G. Mutational analysis of the K+-competitive inhibitor site of gastric H,K-ATPase. *Biochemistry.* 2001;40(25):7480-90.
[23] Munson K, Vagin O, Sachs G, Karlish S. Molecular modeling of SCH28080 binding to the gastric H,K-ATPase and MgATP interactions with SERCA- and Na,K-ATPases. *Ann. N Y Acad. Sci.* 2003;986:106-10.
[24] Shin JM, Kim N. Pharmacokinetics and pharmacodynamics of the proton pump inhibitors. *J. Neurogastroenterol.. Motil.* 2013;19(1):25-35.
[25] Welage LS, Berardi RR. Evaluation of omeprazole, lansoprazole, pantoprazole, and rabeprazole in the treatment of acid-related diseases. *J. Am. Pharm. Assoc.* (Wash). 2000;40(1):52-62; quiz 121-3.

[26] Feldman M, Burton ME. Histamine2-receptor antagonists. Standard therapy for acid-peptic diseases (2). *N. Engl. J. Med.* 1990;323(25):1749-55.

[27] Shin JM, Cho YM, Sachs G. Chemistry of covalent inhibition of the gastric (H+, K+)-ATPase by proton pump inhibitors. *J. Am. Chem. Soc.* 2004;126(25):7800-11.

[28] Shin JM, Sachs G. Restoration of acid secretion following treatment with proton pump inhibitors. *Gastroenterology. 2002*;123(5):1588-97.

[29] Besancon M, Simon A, Sachs G, Shin JM. Sites of reaction of the gastric H,K-ATPase with extracytoplasmic thiol reagents. *J. Biol. Chem.* 1997;272(36):22438-46.

[30] Kromer W. Relative efficacies of gastric proton-pump inhibitors on a milligram basis: desired and undesired SH reactions. Impact of chirality. *Scand. J. Gastroenterol. Suppl.* 2001(234):3-9.

[31] Gerloff J, Mignot A, Barth H, Heintze K. Pharmacokinetics and absolute bioavailability of lansoprazole. *Eur. J. Clin. Pharmacol.* 1996;50(4):293-7.

[32] Radhofer-Welte S. Pharmacokinetics and metabolism of the proton pump inhibitor pantoprazole in man. *Drugs Today* (Barc). 1999;35(10):765-72.

[33] Fuhr U, Jetter A. Rabeprazole: pharmacokinetics and pharmacokinetic drug interactions. *Pharmazie.* 2002;57(9):595-601.

[34] Shin JM, Sachs G. Differences in binding properties of two proton pump inhibitors on the gastric H+,K+-ATPase in vivo. *Biochem. Pharmacol.* 2004;68(11):2117-27.

[35] Krag M, Perner A, Wetterslev J, Moller MH. Stress ulcer prophylaxis in the intensive care unit: is it indicated? A topical systematic review. *Acta Anaesthesiol. Scand. 2013*;57(7):835-47.

[36] Herzig SJ, Rothberg MB, Feinbloom DB, Howell MD, Ho KK, Ngo LH, et al. Risk factors for nosocomial gastrointestinal bleeding and use of acid-suppressive medication in non-critically ill patients. *J. Gen. Intern. Med.* 2013;28(5):683-90.

[37] Cederberg C, Andersson T, Skanberg I. Omeprazole: pharmacokinetics and metabolism in man. *Scand. J. Gastroenterol. Suppl.* 1989;166:33-40; discussion 1-2.

[38] Lind T, Rydberg L, Kyleback A, Jonsson A, Andersson T, Hasselgren G, et al. Esomeprazole provides improved acid control vs. omeprazole In patients with symptoms of gastro-oesophageal reflux disease. *Aliment Pharmacol. Ther.* 2000;14(7):861-7.

[39] Rodvold KA. Clinical pharmacokinetics of clarithromycin. *Clin. Pharmacokinet.* 1999;37(5):385-98.

[40] Cohen S, Booth GH, Jr. Gastric acid secretion and lower-esophageal-sphincter pressure in response to coffee and caffeine. *N. Engl. J. Med.* 1975;293(18):897-9.

[41] Histamine 2-receptor antagonists--standard therapy for acid-peptic diseases. *N. Engl. J. Med.* 1991;324(18):1289-90.

[42] Gillen D, McColl KE. Problems related to acid rebound and tachyphylaxis. *Best Pract. Res. Clin. Gastroenterol.* 2001;15(3):487-95.

[43] Bjornsson E, Abrahamsson H, Simren M, Mattsson N, Jensen C, Agerforz P, et al. Discontinuation of proton pump inhibitors in patients on long-term therapy: a double-blind, placebo-controlled trial. *Aliment Pharmacol. Ther.* 2006;24(6):945-54.

[44] Furuta T, Ohashi K, Kamata T, Takashima M, Kosuge K, Kawasaki T, et al. Effect of genetic differences in omeprazole metabolism on cure rates for Helicobacter pylori infection and peptic ulcer. *Ann Intern Med.* 1998;129(12):1027-30.

[45] Furuta T, Ohashi K, Kosuge K, Zhao XJ, Takashima M, Kimura M, et al. CYP2C19 genotype status and effect of omeprazole on intragastric pH in humans. *Clin. Pharmacol. Ther.* 1999;65(5):552-61.

[46] Tanigawara Y, Aoyama N, Kita T, Shirakawa K, Komada F, Kasuga M, et al. CYP2C19 genotype-related efficacy of omeprazole for the treatment of infection caused by Helicobacter pylori. *Clin. Pharmacol. Ther.* 1999;66(5):528-34.

[47] Kim KA, Park PW, Hong SJ, Park JY. The effect of CYP2C19 polymorphism on the pharmacokinetics and pharmacodynamics of clopidogrel: a possible mechanism for clopidogrel resistance. *Clin. Pharmacol. Ther.* 2008;84(2):236-42.

[48] Farid NA, Payne CD, Small DS, Winters KJ, Ernest CS, 2nd, Brandt JT, et al. Cytochrome P450 3A inhibition by ketoconazole affects prasugrel and clopidogrel pharmacokinetics and pharmacodynamics differently. *Clin. Pharmacol. Ther.* 2007;81(5):735-41.

[49] Li XQ, Andersson TB, Ahlstrom M, Weidolf L. Comparison of inhibitory effects of the proton pump-inhibiting drugs omeprazole, esomeprazole, lansoprazole, pantoprazole, and rabeprazole on human cytochrome P450 activities. *Drug Metab. Dispos.* 2004;32(8):821-7.

[50] Gilard M, Arnaud B, Cornily JC, Le Gal G, Lacut K, Le Calvez G, et al. Influence of omeprazole on the antiplatelet action of clopidogrel associated with aspirin: the randomized, double-blind OCLA

(Omeprazole CLopidogrel Aspirin) study. *J. Am. Coll. Cardiol.* 2008;51(3):256-60.

[51] Ho PM, Maddox TM, Wang L, Fihn SD, Jesse RL, Peterson ED, et al. Risk of adverse outcomes associated with concomitant use of clopidogrel and proton pump inhibitors following acute coronary syndrome. *JAMA*. 2009;301(9):937-44.

[52] Juurlink DN, Gomes T, Ko DT, Szmitko PE, Austin PC, Tu JV, et al. A population-based study of the drug interaction between proton pump inhibitors and clopidogrel. *CMAJ*. 2009;180(7):713-8.

[53] Simon T, Verstuyft C, Mary-Krause M, Quteineh L, Drouet E, Meneveau N, et al. Genetic determinants of response to clopidogrel and cardiovascular events. *N. Engl. J. Med.* 2009;360(4):363-75.

[54] Bhatt DL, Cryer BL, Contant CF, Cohen M, Lanas A, Schnitzer TJ, et al. Clopidogrel with or without omeprazole in coronary artery disease. *N. Engl. J. Med.* 2010;363(20):1909-17.

[55] Abraham NS, Hlatky MA, Antman EM, Bhatt DL, Bjorkman DJ, Clark CB, et al. ACCF/ACG/AHA 2010 expert consensus document on the concomitant use of proton pump inhibitors and thienopyridines: a focused update of the ACCF/ACG/AHA 2008 expert consensus document on reducing the gastrointestinal risks of antiplatelet therapy and NSAID use. A Report of the American College of Cardiology Foundation Task Force on Expert Consensus Documents. *J. Am. Coll Cardiol.* 2010;56(24):2051-66.

[56] Gugler R, Jensen JC. Omeprazole inhibits oxidative drug metabolism. Studies with diazepam and phenytoin in vivo and 7-ethoxycoumarin in vitro. *Gastroenterology*. 1985;89(6):1235-41.

[57] Ishizaki T, Chiba K, Manabe K, Koyama E, Hayashi M, Yasuda S, et al. Comparison of the interaction potential of a new proton pump inhibitor, E3810, versus omeprazole with diazepam in extensive and poor metabolizers of S-mephenytoin 4'-hydroxylation. *Clin. Pharmacol. Ther.* 1995;58(2):155-64.

[58] Karol MD, Locke CS, Cavanaugh JH. Lack of pharmacokinetic interaction between lansoprazole and intravenously administered phenytoin. *J. Clin. Pharmacol.* 1999;39(12):1283-9.

[59] Shi S, Klotz U. Proton pump inhibitors: an update of their clinical use and pharmacokinetics. *Eur. J. Clin. Pharmacol.* 2008;64(10):935-51.

[60] Svensson US, Ashton M, Trinh NH, Bertilsson L, Dinh XH, Nguyen VH, et al. Artemisinin induces omeprazole metabolism in human beings. *Clin. Pharmacol. Ther.* 1998;64(2):160-7.

[61] Yin OQ, Tomlinson B, Waye MM, Chow AH, Chow MS. Pharmacogenetics and herb-drug interactions: experience with Ginkgo biloba and omeprazole. *Pharmacogenetics.* 2004;14(12):841-50.

[62] Abbvie. http://www.rxabbvie.com/pdf/viekirapak_pi.pdf.Accessed February 2015.

[63] Information P. Harvoni (ledipasvir-sofosbuvir). Gilead Sciences, Foster City, CA;Accessed February 2015.

[64] Eriksson S, Langstrom G, Rikner L, Carlsson R, Naesdal J. Omeprazole and H2-receptor antagonists in the acute treatment of duodenal ulcer, gastric ulcer and reflux oesophagitis: a meta-analysis. *Eur. J. Gastroenterol. Hepatol.* 1995;7(5):467-75.

[65] Poynard T, Lemaire M, Agostini H. Meta-analysis of randomized clinical trials comparing lansoprazole with ranitidine or famotidine in the treatment of acute duodenal ulcer. *Eur. J. Gastroenterol. Hepatol.* 1995;7(7):661-5.

[66] Dekkers CP, Beker JA, Thjodleifsson B, Gabryelewicz A, Bell NE, Humphries TJ. Comparison of rabeprazole 20 mg versus omeprazole 20 mg in the treatment of active duodenal ulcer: a European multicentre study. *Aliment Pharmacol. Ther.* 1999;13(2):179-86.

[67] Bader JP, Delchier JC. Clinical efficacy of pantoprazole compared with ranitidine. *Aliment Pharmacol. Ther.* 1994;8 Suppl 1:47-52.

[68] Allison MC, Howatson AG, Torrance CJ, Lee FD, Russell RI. Gastrointestinal damage associated with the use of nonsteroidal antiinflammatory drugs. *N. Engl. J. Med.* 1992;327(11):749-54.

[69] Whittle BJ. Mechanisms underlying gastric mucosal damage induced by indomethacin and bile-salts, and the actions of prostaglandins. *Br. J. Pharmacol.* 1977;60(3):455-60.

[70] Hawkey CJ, Karrasch JA, Szczepanski L, Walker DG, Barkun A, Swannell AJ, et al. Omeprazole compared with misoprostol for ulcers associated with nonsteroidal antiinflammatory drugs. Omeprazole versus Misoprostol for NSAID-induced Ulcer Management (OMNIUM) Study Group. *N. Engl. J. Med.* 1998;338(11):727-34.

[71] Yeomans ND, Tulassay Z, Juhasz L, Racz I, Howard JM, van Rensburg CJ, et al. A comparison of omeprazole with ranitidine for ulcers associated with nonsteroidal antiinflammatory drugs. Acid Suppression Trial: Ranitidine versus Omeprazole for NSAID-associated Ulcer Treatment (ASTRONAUT) Study Group. *N. Engl. J. Med.* 1998;338(11):719-26.

[72] Agrawal NM, Campbell DR, Safdi MA, Lukasik NL, Huang B, Haber MM. Superiority of lansoprazole vs ranitidine in healing nonsteroidal anti-inflammatory drug-associated gastric ulcers: results of a double-blind, randomized, multicenter study. NSAID-Associated Gastric Ulcer Study Group. *Arch. Intern. Med.* 2000;160(10):1455-61.

[73] Scheiman JM, Yeomans ND, Talley NJ, Vakil N, Chan FK, Tulassay Z, et al. Prevention of ulcers by esomeprazole in at-risk patients using non-selective NSAIDs and COX-2 inhibitors. *Am. J. Gastroenterol.* 2006;101(4):701-10.

[74] Scheiman JM, Devereaux PJ, Herlitz J, Katelaris PH, Lanas A, Veldhuyzen van Zanten S, et al. Prevention of peptic ulcers with esomeprazole in patients at risk of ulcer development treated with low-dose acetylsalicylic acid: a randomised, controlled trial (OBERON). *Heart.* 2011;97(10):797-802.

[75] Leontiadis GI, Sharma VK, Howden CW. Systematic review and meta-analysis of proton pump inhibitor therapy in peptic ulcer bleeding. *BMJ.* 2005;330(7491):568.

[76] Lin HJ, Lo WC, Lee FY, Perng CL, Tseng GY. A prospective randomized comparative trial showing that omeprazole prevents rebleeding in patients with bleeding peptic ulcer after successful endoscopic therapy. *Arch. Intern. Med.* 1998;158(1):54-8.

[77] Collins R, Langman M. Treatment with histamine H2 antagonists in acute upper gastrointestinal hemorrhage. Implications of randomized trials. *N. Engl. J. Med.* 1985;313(11):660-6.

[78] Walt RP, Cottrell J, Mann SG, Freemantle NP, Langman MJ. Continuous intravenous famotidine for haemorrhage from peptic ulcer. *Lancet.* 1992;340(8827):1058-62.

[79] Levine JE, Leontiadis GI, Sharma VK, Howden CW. Meta-analysis: the efficacy of intravenous H2-receptor antagonists in bleeding peptic ulcer. *Aliment Pharmacol. Ther.* 2002;16(6):1137-42.

[80] van Rensburg C, Barkun AN, Racz I, Fedorak R, Bornman PC, Beglinger C, et al. Clinical trial: intravenous pantoprazole vs. ranitidine for the prevention of peptic ulcer rebleeding: a multicentre, multinational, randomized trial. *Aliment Pharmacol. Ther.* 2009;29(5):497-507.

[81] Zed PJ, Loewen PS, Slavik RS, Marra CA. Meta-analysis of proton pump inhibitors in treatment of bleeding peptic ulcers. *Ann. Pharmacother.* 2001;35(12):1528-34.

[82] Oviedo JA, Wolfe WM. Management of stress related erosive syndrome. In: TM Bayless, A Diehl (eds), *Advanced Therapy in Gastroenterolgoy and Liver Disease*. 2005;Michigan: BC Decker:161-66.
[83] Wang CH, Ma MH, Chou HC, Yen ZS, Yang CW, Fang CC, et al. High-dose vs non-high-dose proton pump inhibitors after endoscopic treatment in patients with bleeding peptic ulcer: a systematic review and meta-analysis of randomized controlled trials. *Arch. Intern. Med.* 2010;170(9):751-8.
[84] Neumann I, Letelier LM, Rada G, Claro JC, Martin J, Howden CW, et al. Comparison of different regimens of proton pump inhibitors for acute peptic ulcer bleeding. *Cochrane Database Syst Rev*. 2013;6:CD007999.
[85] Sachar H, Vaidya K, Laine L. Intermittent vs continuous proton pump inhibitor therapy for high-risk bleeding ulcers: a systematic review and meta-analysis. *JAMA Intern Med*. 2014;174(11):1755-62.
[86] Laine L, Shah A, Bemanian S. Intragastric pH with oral vs intravenous bolus plus infusion proton-pump inhibitor therapy in patients with bleeding ulcers. *Gastroenterology.* 2008;134(7):1836-41.
[87] Tsoi KK, Hirai HW, Sung JJ. Meta-analysis: comparison of oral vs. intravenous proton pump inhibitors in patients with peptic ulcer bleeding. *Aliment Pharmacol. Ther*. 2013;38(7):721-8.
[88] Sung JJ, Suen BY, Wu JC, Lau JY, Ching JY, Lee VW, et al. Effects of intravenous and oral esomeprazole in the prevention of recurrent bleeding from peptic ulcers after endoscopic therapy. *Am. J. Gastroenterol.* 2014;109(7):1005-10.
[89] Spirt MJ. Stress-related mucosal disease: risk factors and prophylactic therapy. *Clin. Ther*. 2004;26(2):197-213.
[90] Spirt MJ, Stanley S. Update on stress ulcer prophylaxis in critically ill patients. *Crit. Care Nurse*. 2006;26(1):18-20, 2-8; quiz 9.
[91] Cook DJ, Fuller HD, Guyatt GH, Marshall JC, Leasa D, Hall R, et al. Risk factors for gastrointestinal bleeding in critically ill patients. Canadian Critical Care Trials Group. *N. Engl. J. Med.* 1994;330(6):377-81.
[92] Ben-Menachem T, Fogel R, Patel RV, Touchette M, Zarowitz BJ, Hadzijahic N, et al. Prophylaxis for stress-related gastric hemorrhage in the medical intensive care unit. A randomized, controlled, single-blind study. *Ann. Intern. Med.* 1994;121(8):568-75.
[93] Shuman RB, Schuster DP, Zuckerman GR. Prophylactic therapy for stress ulcer bleeding: a reappraisal. *Ann. Intern. Med.* 1987;106(4):562-7.

[94] Cook DJ, Griffith LE, Walter SD, Guyatt GH, Meade MO, Heyland DK, et al. The attributable mortality and length of intensive care unit stay of clinically important gastrointestinal bleeding in critically ill patients. *Crit. Care.* 2001;5(6):368-75.

[95] ASHP Therapeutic Guidelines on Stress Ulcer Prophylaxis. ASHP Commission on Therapeutics and approved by the ASHP Board of Directors on November 14, 1998. *Am. J. Health Syst. Pharm.* 1999;56(4):347-79.

[96] Barkun AN, Bardou M, Pham CQ, Martel M. Proton pump inhibitors vs. histamine 2 receptor antagonists for stress-related mucosal bleeding prophylaxis in critically ill patients: a meta-analysis. *Am. J. Gastroenterol. 2012*;107(4):507-20; quiz 21.

[97] Barkun AN, Adam V, Martel M, Bardou M. Cost-effectiveness analysis: stress ulcer bleeding prophylaxis with proton pump inhibitors, H2 receptor antagonists. *Value Health.* 2013;16(1):14-22.

[98] Wahlqvist P, Reilly MC, Barkun A. Systematic review: the impact of gastro-oesophageal reflux disease on work productivity. *Aliment Pharmacol. Ther.* 2006;24(2):259-72.

[99] Shaheen NJ, Hansen RA, Morgan DR, Gangarosa LM, Ringel Y, Thiny MT, et al. The burden of gastrointestinal and liver diseases, 2006. *Am. J. Gastroenterol.* 2006;101(9):2128-38.

[100] Locke GR, 3rd, Talley NJ, Fett SL, Zinsmeister AR, Melton LJ, 3rd. Prevalence and clinical spectrum of gastroesophageal reflux: a population-based study in Olmsted County, Minnesota. *Gastroenterology.* 1997;112(5):1448-56.

[101] Sheikh I, Waghray A, Waghray N, Dong C, Wolfe MM. Consumer use of over-the-counter proton pump inhibitors in patients with gastroesophageal reflux disease. *Am. J. Gastroenterol.* 2014;109(6):789-94.

[102] Wolfe MM, Lowe RC. Investing in the Future of GERD. *Journal of Clinical Gastroenterology.* 2007;41(Supp 2):S209-16.

[103] Chiba N, De Gara CJ, Wilkinson JM, Hunt RH. Speed of healing and symptom relief in grade II to IV gastroesophageal reflux disease: a meta-analysis. *Gastroenterology.* 1997;112(6):1798-810.

[104] Donnellan C, Sharma N, Preston C, Moayyedi P. Medical treatments for the maintenance therapy of reflux oesophagitis and endoscopic negative reflux disease. *Cochrane Database Syst. Rev.* 2005(2):CD003245.

[105] Jansen JB, Van Oene JC. Standard-dose lansoprazole is more effective than high-dose ranitidine in achieving endoscopic healing and symptom

relief in patients with moderately severe reflux oesophagitis. The Dutch Lansoprazole Study Group. *Aliment Pharmacol. Ther.* 1999;13(12):1611-20.

[106] Southworth MR, Temple R. Interaction of clopidogrel and omeprazole. *N. Engl. J. Med.* 2010;363(20):1977.

[107] Bardhan KD, Morris P, Thompson M, Dhande DS, Hinchliffe RF, Jones RB, et al. Omeprazole in the treatment of erosive oesophagitis refractory to high dose cimetidine and ranitidine. *Gut.* 1990;31(7):745-9.

[108] McDonagh MS, Carson S, Thakurta S. Drug Class Review: Proton Pump Inhibitors: Final Report Update 5. Portland (OR); 2009.

[109] Kahrilas PJ, Falk GW, Johnson DA, Schmitt C, Collins DW, Whipple J, et al. Esomeprazole improves healing and symptom resolution as compared with omeprazole in reflux oesophagitis patients: a randomized controlled trial. The Esomeprazole Study Investigators. *Aliment Pharmacol. Ther.* 2000;14(10):1249-58.

[110] Richter JE, Kahrilas PJ, Johanson J, Maton P, Breiter JR, Hwang C, et al. Efficacy and safety of esomeprazole compared with omeprazole in GERD patients with erosive esophagitis: a randomized controlled trial. *Am. J. Gastroenterol.* 2001;96(3):656-65.

[111] Castell DO, Kahrilas PJ, Richter JE, Vakil NB, Johnson DA, Zuckerman S, et al. Esomeprazole (40 mg) compared with lansoprazole (30 mg) in the treatment of erosive esophagitis. *Am. J. Gastroenterol.* 2002;97(3):575-83.

[112] Fass R, Sontag SJ, Traxler B, Sostek M. Treatment of patients with persistent heartburn symptoms: a double-blind, randomized trial. *Clin Gastroenterol Hepatol.* 2006;4(1):50-6.

[113] Swarbrick ET, Gough AL, Foster CS, Christian J, Garrett AD, Langworthy CH. Prevention of recurrence of oesophageal stricture, a comparison of lansoprazole and high-dose ranitidine. *Eur. J. Gastroenterol. Hepatol.* 1996;8(5):431-8.

[114] Marks RD, Richter JE, Rizzo J, Koehler RE, Spenney JG, Mills TP, et al. Omeprazole versus H2-receptor antagonists in treating patients with peptic stricture and esophagitis. *Gastroenterology.* 1994;106(4):907-15.

[115] Hetzel DJ, Dent J, Reed WD, Narielvala FM, Mackinnon M, McCarthy JH, et al. Healing and relapse of severe peptic esophagitis after treatment with omeprazole. *Gastroenterology.* 1988;95(4):903-12.

[116] Vigneri S, Termini R, Leandro G, Badalamenti S, Pantalena M, Savarino V, et al. A comparison of five maintenance therapies for reflux esophagitis. *N. Engl. J. Med.* 1995;333(17):1106-10.

[117] Gunaratnam NT, Jessup TP, Inadomi J, Lascewski DP. Sub-optimal proton pump inhibitor dosing is prevalent in patients with poorly controlled gastro-oesophageal reflux disease. *Aliment Pharmacol. Ther.* 2006;23(10):1473-7.

[118] Li H, Helander HF. Hypergastrinemia increases proliferation of gastroduodenal epithelium during gastric ulcer healing in rats. *Dig. Dis. Sci.* 1996;41(1):40-8.

[119] Malagelada JR, Edis AJ, Adson MA, van Heerden JA, Go VL. Medical and surgical options in the management of patients with gastrinoma. *Gastroenterology.* 1983;84(6):1524-32.

[120] Howard JM, Chremos AN, Collen MJ, McArthur KE, Cherner JA, Maton PN, et al. Famotidine, a new, potent, long-acting histamine H2-receptor antagonist: comparison with cimetidine and ranitidine in the treatment of Zollinger-Ellison syndrome. *Gastroenterology.* 1985;88(4):1026-33.

[121] Maton PN. Role of acid suppressants in patients with Zollinger-Ellison syndrome. *Aliment Pharmacol. Ther.* 1991;5 Suppl 1:25-35.

[122] Metz DC, Strader DB, Orbuch M, Koviack PD, Feigenbaum KM, Jensen RT. Use of omeprazole in Zollinger-Ellison syndrome: a prospective nine-year study of efficacy and safety. *Aliment Pharmacol. Ther.* 1993;7(6):597-610.

[123] Metz DC, Pisegna JR, Fishbeyn VA, Benya RV, Feigenbaum KM, Koviack PD, et al. Currently used doses of omeprazole in Zollinger-Ellison syndrome are too high. *Gastroenterology.* 1992;103(5):1498-508.

[124] Hirschowitz BI, Simmons J, Mohnen J. Clinical outcome using lansoprazole in acid hypersecretors with and without Zollinger-Ellison syndrome: a 13-year prospective study. *Clin. Gastroenterol. Hepatol.* 2005;3(1):39-48.

[125] Pospai D, Cadiot G, Forestier S, Ruszniewski P, Coste T, Escourrou J, et al. [Effectiveness and safety of lansoprazole in the treatment of Zollinger-Ellison syndrome. First six months of treatment]. *Gastroenterol. Clin. Biol.* 1998;22(10):801-8.

[126] Wolfe MM, Jensen RT. Zollinger-Ellison syndrome. Current concepts in diagnosis and management. *N. Engl. J. Med.* 1987;317(19):1200-9.

[127] Leonard J, Marshall JK, Moayyedi P. Systematic review of the risk of enteric infection in patients taking acid suppression. *Am. J. Gastroenterol.* 2007;102(9):2047-56; quiz 57.

[128] Kwok CS, Arthur AK, Anibueze CI, Singh S, Cavallazzi R, Loke YK. Risk of Clostridium difficile infection with acid suppressing drugs and antibiotics: meta-analysis. *Am. J. Gastroenterol.* 2012;107(7):1011-9.
[129] Janarthanan S, Ditah I, Adler DG, Ehrinpreis MN. Clostridium difficile-associated diarrhea and proton pump inhibitor therapy: a meta-analysis. *Am. J. Gastroenterol.* 2012;107(7):1001-10.
[130] FDA Drug Safety Communication: Clostridium difficile-associated diarrhea can be associated with stomach acid drugs known as proton pump inhibitors (PPIs). Issued 02-08-2012. Accessed 2/2015.
[131] Gulmez SE, Holm A, Frederiksen H, Jensen TG, Pedersen C, Hallas J. Use of proton pump inhibitors and the risk of community-acquired pneumonia: a population-based case-control study. *Arch. Intern. Med.* 2007;167(9):950-5.
[132] Laheij RJ, Sturkenboom MC, Hassing RJ, Dieleman J, Stricker BH, Jansen JB. Risk of community-acquired pneumonia and use of gastric acid-suppressive drugs. *JAMA*. 2004;292(16):1955-60.
[133] Sarkar M, Hennessy S, Yang YX. Proton-pump inhibitor use and the risk for community-acquired pneumonia. *Ann. Intern. Med.* 2008;149(6):391-8.
[134] Eom CS, Jeon CY, Lim JW, Cho EG, Park SM, Lee KS. Use of acid-suppressive drugs and risk of pneumonia: a systematic review and meta-analysis. *CMAJ.* 2011;183(3):310-9.
[135] Jena AB, Sun E, Goldman DP. Confounding in the association of proton pump inhibitor use with risk of community-acquired pneumonia. *J. Gen. Intern. Med.* 2013;28(2):223-30.
[136] Freedberg DE, Abrams JA. Does confounding explain the association between PPIs and Clostridium difficile-related diarrhea? *Am. J. Gastroenterol.* 2013;108(2):278-9.
[137] Filion KB, Chateau D, Targownik LE, Gershon A, Durand M, Tamim H, et al. Proton pump inhibitors and the risk of hospitalisation for community-acquired pneumonia: replicated cohort studies with meta-analysis. *Gut.* 2014;63(4):552-8.
[138] McColl KE. Effect of proton pump inhibitors on vitamins and iron. *Am. J. Gastroenterol.* 2009;104 Suppl 2:S5-9.
[139] Recker RR. Calcium absorption and achlorhydria. *N. Engl. J. Med.* 1985;313(2):70-3.
[140] Mizunashi K, Furukawa Y, Katano K, Abe K. Effect of omeprazole, an inhibitor of H+,K(+)-ATPase, on bone resorption in humans. *Calcif. Tissue Int.* 1993;53(1):21-5.

[141] Yang YX, Lewis JD, Epstein S, Metz DC. Long-term proton pump inhibitor therapy and risk of hip fracture. *JAMA*. 2006;296(24):2947-53.

[142] Corley DA, Kubo A, Zhao W, Quesenberry C. Proton pump inhibitors and histamine-2 receptor antagonists are associated with hip fractures among at-risk patients. *Gastroenterology*. 2010;139(1):93-101.

[143] Targownik LE, Lix LM, Leung S, Leslie WD. Proton-pump inhibitor use is not associated with osteoporosis or accelerated bone mineral density loss. *Gastroenterology*. 2010;138(3):896-904.

[144] Targownik LE, Leslie WD, Davison KS, Goltzman D, Jamal SA, Kreiger N, et al. The relationship between proton pump inhibitor use and longitudinal change in bone mineral density: a population-based study [corrected] from the Canadian Multicentre Osteoporosis Study (CaMos). *Am. J. Gastroenterol*. 2012;107(9):1361-9.

[145] Yu EW, Bauer SR, Bain PA, Bauer DC. Proton pump inhibitors and risk of fractures: a meta-analysis of 11 international studies. *Am. J. Med*. 2011;124(6):519-26.

[146] Khalili H, Huang ES, Jacobson BC, Camargo CA, Jr., Feskanich D, Chan AT. Use of proton pump inhibitors and risk of hip fracture in relation to dietary and lifestyle factors: a prospective cohort study. *BMJ*. 2012;344:e372.

[147] FDA Drug Safety Communication: Possible increased risk of fractures of the hip, wrist, and spine with the use of proton pump inhibitors. Issued 05-25-2010. Accessed 02/2015.

[148] Hess MW, Hoenderop JG, Bindels RJ, Drenth JP. Systematic review: hypomagnesaemia induced by proton pump inhibition. *Aliment Pharmacol. Ther*. 2012;36(5):405-13.

[149] Axelson J, Hakanson R, Rosengren E, Sundler F. Hypergastrinaemia induced by acid blockade evokes enterochromaffin-like (ECL) cell hyperplasia in chicken, hamster and guinea-pig stomach. *Cell Tissue Res*. 1988;254(3):511-6.

[150] Freston JW. Omeprazole, hypergastrinemia, and gastric carcinoid tumors. *Ann. Intern. Med*. 1994;121(3):232-3.

[151] van Soest EM, van Rossum LG, Dieleman JP, van Oijen MG, Siersema PD, Sturkenboom MC, et al. Proton pump inhibitors and the risk of colorectal cancer. *Am. J. Gastroenterol*. 2008;103(4):966-73.

[152] Bajaj JS, Zadvornova Y, Heuman DM, Hafeezullah M, Hoffmann RG, Sanyal AJ, et al. Association of proton pump inhibitor therapy with spontaneous bacterial peritonitis in cirrhotic patients with ascites. *Am. J. Gastroenterol*. 2009;104(5):1130-4.

[153] Kwon JH, Koh SJ, Kim W, Jung YJ, Kim JW, Kim BG, et al. Mortality associated with proton pump inhibitors in cirrhotic patients with spontaneous bacterial peritonitis. *J. Gastroenterol. Hepatol.* 2014;29(4):775-81.

[154] Ratelle M, Perreault S, Villeneuve JP, Tremblay L. Association between proton pump inhibitor use and spontaneous bacterial peritonitis in cirrhotic patients with ascites. *Can. J. Gastroenterol. Hepatol.* 2014;28(6):330-4.

[155] Harmark L, van der Wiel HE, de Groot MC, van Grootheest AC. Proton pump inhibitor-induced acute interstitial nephritis. *Br. J. Clin. Pharmacol.*. 2007;64(6):819-23.

[156] Ray S, Delaney M, Muller AF. Proton pump inhibitors and acute interstitial nephritis. *BMJ.* 2010;341:c4412.

[157] Choudhry U, Boyce HW, Jr., Coppola D. Proton pump inhibitor-associated gastric polyps: a retrospective analysis of their frequency, and endoscopic, histologic, and ultrastructural characteristics. *Am. J. Clin. Pathol.* 1998;110(5):615-21.

In: Proton Pump Inhibitors (PPIs)
Editor: Barbara Parker
ISBN: 978-1-63482-890-1

Chapter II

Benefits and Risks of Routine Use of Proton Pump Inhibitors at the Intensive Care Unit

Lukas Buendgens and Frank Tacke *
Dept. of Medicine III, University Hospital Aachen, Germany

Abstract

Stress-related mucosal disease (SRMD) is present in nearly all critical ill patients. SRMD leads to an elevated risk of upper gastrointestinal (GI) bleedings, a potentially life-threatening complication associated with a mortality rate of about 30%. This led to the recommendation for stress ulcer prophylaxis in critically ill patients, especially to those with risk factors for GI bleeding such as mechanical ventilation or coagulopathies. Proton pump inhibitors (PPI) effectively prevent gastrointestinal bleedings in critically ill patients at the intensive care unit (ICU). Large meta-analyses including up to 1720 patients from 14 clinical trials revealed that PPI seem to be more effective than histamine 2 receptor antagonists (H2RA) in preventing clinically significant upper GI bleedings in critically ill patients, although they did

* Address for correspondence:Prof. Dr. Frank Tacke, MD, PhD; University Hospital Aachen, Dept of Medicine III; Pauwelsstr 30, D-52074 Aachen, Germany; Tel. +49-241-80-35848, Fax +49-241-80-82455; Email: frank.tacke@gmx.net

not reduce overall mortality in ICU patients. Moreover, recent studies revealed that the routine use of PPI at the ICU can be associated with adverse events such as a significantly increased risk of infectious complications, especially of nosocomial pneumonia and *Clostridium difficile*–associated diarrhea (CDAD). Likewise, PPI can be toxic for both the liver and the bone marrow, and some PPI show clinically relevant interactions with important other drugs like clopidogrel. Therefore, the agent of choice, the specific balance of risks and benefits for individual patients as well as the possible dose of PPI has to be chosen carefully. Alternatives to PPI prophylaxis include histamine receptor blockers and/or sucralfate. Instead of routine PPI use for bleeding prophylaxis, further trials should investigate risk-adjusted algorithms, balancing benefits and threats of PPI medication at the ICU.

Keywords: Critical illness, sepsis, stress ulcer, GI bleeding, pneumonia, clostrium difficile colitis

Introduction

Stress-related mucosal disease (SRMD) is present in nearly all critical ill patients [1]. Not all of these lesions are clinically relevant, but may lead to an overt or major bleeding complication. Earlier trials reported an incidence of bleeding events in ICU patients of 17% for the period between 1980 and 1998 [2]. Nowadays, this incidence has remarkably decreased, as more recent studies revealed a rate of severe gastrointestinal (GI) bleedings below 1% at the ICU [3-5]. This dramatic decrease in clinically relevant GI hemorrhages might be due to improved care, pharmacological prophylaxis and increased use of enteral feeding [6].

Two pathogenic factors have been proposed to determine the development of SRMD. One is the reduced mucosal blood flow and consecutive local ischemia in critically ill patients [7, 8]. This is not only constricted to patients with instable hemodynamics, such as hypotension and/or vasopressor therapy, but can even develop in patients with normal systemic blood pressure if the splanchnic blood flow is impaired, for example, in mechanically ventilated patients with moderate to high positive end-expiratory pressure (PEEP). On the level of molecular mechanisms, a reduction in local NO synthesis and an up-regulation of endothelin-1 as a potent endogenous vasoconstrictor promote the damage to the mucosa [9, 10]. However, in animal models, ischemia and reperfusion are only able to cause minimal mucosal damage, if no additional damaging factor is present. In fact, if acid is added to the model of ischemia/reperfusion injury of the stomach, the damage increases tenfold [10]. This critical contribution of acid to the progression from mucosal vulnerability to clinically overt mucosal damage provides the rationale behind acid-suppressing drugs in preventing GI hemorrhages related to SRMD (Figure 1).

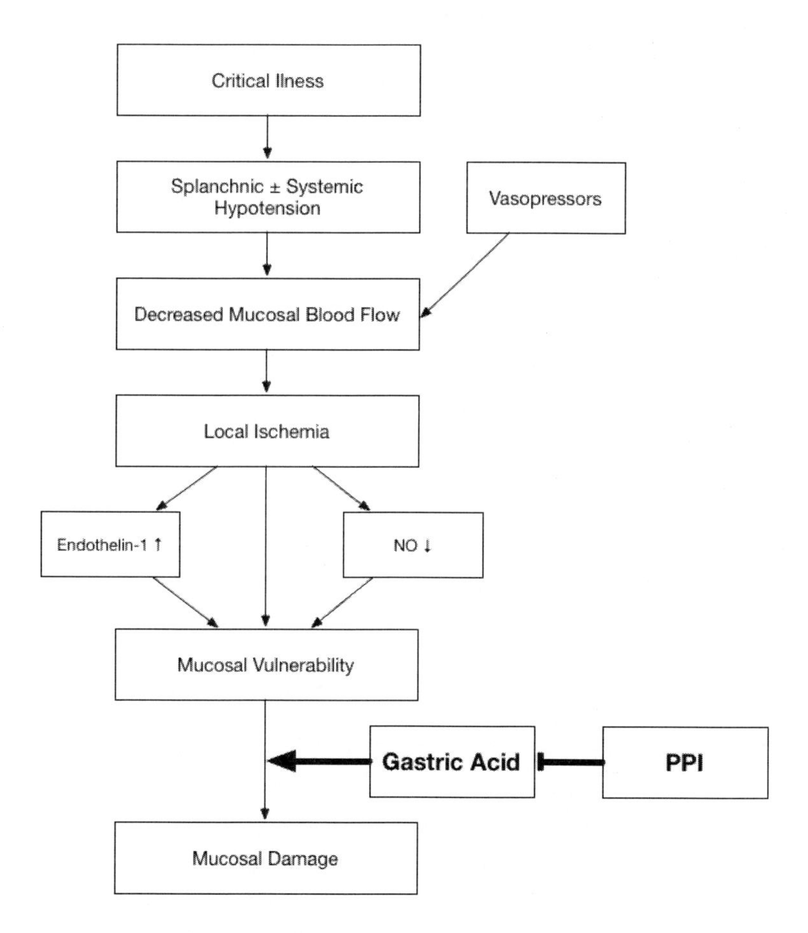

Figure 1. Pathogenesis of stress-related mucosal disease (SRMD) and rationale for the routine use of PPI at the ICU. Abbreviations: ICU = intensive care unit; NO = nitric oxide; PPI = proton pump inhibitor(s).

Major independent risk factors for GI bleeding related to SRMD are especially the need for mechanical ventilation (odds ratio (OR) 15.6) and the presence of coagulopathy (OR 4.3) [12], both increasing the incidence of hemorrhages massively in ICU patients. In addition, specific subgroups such as patients with severe head trauma or burns and those after extended surgeries with operation time exceeding 4h have a significantly increased risk for bleedings. Other, but less well validated risk factors include acute kidney failure, age and male sex [13].

GI bleedings represent a potentially life-threatening complication to critically ill patients. GI bleeding from a stress-related lesion is associated with a significantly higher mortality rate, increasing the relative risk (RR) of death up to 4 [14]. Therefore, pharmacological prophylaxis of SRMD-related bleeding in high-risk patients is nowadays an established practice on the ICU in severely ill patients.

Indications for Pharmacological Prophylaxis of SRMD-Related Hemorrhages

There are explicit guidelines for the use of stress ulcer prophylaxis to prevent upper gastrointestinal bleeding for ICU patients with sepsis and septic shock [15, 16], and these recommendations are oftentimes extrapolated to general medical and surgical ICU populations. Due to the clinical impact of a major gastrointestinal bleeding both international and national guidelines recommend the use of stress ulcer prophylaxis, but only in patients with risk factors (i.e. mechanical ventilation >48h, coagulopathy, hypotension). For patients without those risk factors, there is a recommendation against a prophylaxis [15, 16]. In our unit, we recommend risk stratification for ulcer prophylaxis. Critically ill patients with one of the following risk factors should (strong evidence) receive an ulcer prophylaxis: mechanical ventilation, coagulopathy, history of an upper gastrointestinal bleeding within the past 12 months, severe sepsis or septic shock, or cardiogenic shock. For the following, we recommend ulcer prophylaxis (weaker evidence): burn patients, craniocerebral injury, acute renal failure, known peptic ulcer disease, patients after kidney or liver transplantation, intake of non-steroidal anti-inflammatory drugs (NSAID) or requiring high-dose glucocorticoids. However, it is mandatory to frequently re-evaluate the individual indication both during and after the stay at the ICU.

Those recommendations are based on several meta-analyses, which clearly show a significant reduction in bleeding rates, if a pharmacological prophylaxis is used [17-20]. The odds ratios for clinically relevant bleedings range between 0.44 and 0.72, if a stress ulcer prophylaxis is used. However, it is important to emphasize that the reduced risk of bleeding does not directly translate into a benefit regarding mortality.

There is no evidence for stress ulcer prophylaxis outside an ICU setting. It is a common concern that stress ulcer prophylaxis is continued even after discharge from the hospital without reevaluation of its indication. This practice causes unnecessary risks (pneumonia, CDAD) as well as unnecessary costs [20].

Role of PPI in the Prophylaxis of SRMD-Related Hemorrhages at the ICU and in the Treatment of Patients with Gastrointestinal Bleedings

There are mainly three pharmacological options for stress ulcer prophylaxis at the ICU: proton pump inhibitors (PPI) and H2-receptor antagonists (H2RA) as acid suppressing drugs and the local mucosa-protective agent sucralfate.

The latter is a valid option and oftentimes used in addition to acid suppression [5]. However, as a single therapy sucralfate was inferior to H2RA in a large multicenter study [22].

Many studies have compared the use of PPI versus H2RA in different critically ill patient populations (summarized in [3]). With respect to the reduction of GI bleedings, the majority of recent meta-analyses demonstrated that PPI are more effective than H2RA (Table 1) [3, 23]. Although a fairly large number of patients is summarized in the recent meta-analyses, all analyses are hampered by the fact that some of the underlying individual studies do not meet highest quality criteria[6]. Importantly, the meta-analyses did not reveal a risk reduction for the overall mortality rate in ICU patients, neither for PPI nor H2RA. Nonetheless, PPI are now considered the agents of first choice for stress ulcer prophylaxis at the ICU [15, 16].

In patients *with* (non-variceal) upper GI bleedings, PPI are undoubtedly the first choice for effective acid suppression. Current guidelines recommend high-dose PPI both before and after endoscopic intervention [25].

PPI are available in both intravenous and oral formulations and can be administered continuously as well as intermittent. There are no direct comparisons between different PPIs in the setting of stress ulcer prophylaxis at the ICU. The most common agents include pantoprazole (daily dose at the ICU 40 mg), omeprazole (daily dose 40mg) and esomeprazole (daily dose 40mg). In case of active GI bleedings, intermittent high dose PPI (e.g., pantoprazole i.v. 40mg twice daily) is efficient in preventing re-bleeding; continuous PPI therapy (e.g., pantoprazole i.v. 240mg/d) is not superior to intermittent administration [26]. All PPIs need to be dose-adjusted in case of concomitant renal or hepatic failure.

Table 1. Meta-analyses comparing efficacy of H2RA vs PPI in ICU patients

Meta-analysis	n	Risk Reduction (Bleeding)	Risk Reduction (Mortality)
Alhazzani et al. 2013 [3]	1720	RR 0.36 (95%-CI 0.19–0.67)	RR 1.01 (95%-CI 0.83–1.24)
Pongprasobchai et al. 2009 [23]	569	OR 0.42 (95%-CI 0.20-0.91)	n/a
Barkun et al. 2012 [24]	1587	OR 0.30 (95%-CI 0.17–0.54)	OR 1.19 (95%-CI 0.84–1.68)
Lin et al. 2010 [1]	936	RD -0.04 (95%-CI −0.09-0.01)	RD 0.00 (95%-CI −0.04-0.05, n.s.)

Abbreviations: ICU = intensive care unit; n/a = not assessed; n = patients included in the meta-analysis; OR = odds ratio; PPI = proton pump inhibitor(s); RD = risk difference.

Adverse Events and Risks of PPI in the Prophylaxis of SRMD-Related Hemorrhages at the ICU

PPI are generally well-tolerated drugs. However, possible side effects include headache, nausea and abdominal pain. In addition, very efficient acid suppression may reduce the physiological barrier to ingested bacteria and can subsequently be associated with gastric and duodenal bacterial overgrowth

[27]. Furthermore, experimental data suggest that the leukocyte function may be impaired in patients taking PPI due to an inhibition of phagocytosis and acidification of the phagolysosome [28, 29].

These mechanisms possibly render the patients receiving PPI susceptible to infections. In consequence, two major infectious complications have to be considered: nosocomial pneumonia and *Clostridium difficile*–associated diarrhea. For patients outside an ICU it has been clearly shown that PPI increase the risk of pneumonia [30-32] as well as the risk for *Clostridium difficile*–associated diarrhea [33]. In the ICU studies comparing PPI or H2RA for stress ulcer prophylaxis (see Table 1), no clear difference for the risk of pneumonia was noted [3]. An earlier meta-analysis comparing different acid suppressing regimen demonstrated a trend towards increased risk of pneumonia in H2RA treated ICU patients compared to no prophylaxis without reaching statistical significance [18]. The risk for *Clostridium difficile*–associated diarrhea has not been comprehensively assessed in most of the prospective trials evaluating the efficacy of PPI-based stress ulcer prophylaxis.

Therefore, we have conducted a large retrospective analysis on our medical ICU, comprising 3286 critically ill patients between 1999 and 2010 [5]. On our unit, 91.3% of patients received stress ulcer prophylaxis either by PPI (55.6%), H2RA (5.8%), sucralfate (10.1%), or combinations (19.8%). Overall, the rate of GI bleeding was very low with only 29 (0.9%) of 3286 patients developing GI hemorrhages during the course of ICU treatment. Also in our retrospective analysis, PPI did not represent an independent risk factor for nosocomial pneumonia. Importantly, 110 patients (3.3%) developed *Clostridium difficile*–associated diarrhea during the course of ICU treatment. PPI administration was identified as an independent risk factor (OR 3.11) for developing *Clostridium difficile*–associated diarrhea at the ICU by multivariate analysis [5] similar to the use of certain antibiotics like cephalosporines and fluoroquinolones that predispose to *Clostridium difficile*–associated diarrhea (Table 2).

Another concern are potential drug-drug-interactions, especially in ICU patients that require multiple drugs and have impaired drug metabolization pathways. Of special interest is a possible interaction between the antiplatelet agent clopidogrel and PPIs. A startling study showed increased cardiovascular events in patients taking both clopidogrel and PPI, which even prompted a warning by the US Food & Drug Administration [34].

Clopidogrel is a prodrug, which needs to be activated in the liver by the enzyme CYP2C19. *In vitro* data indicated that PPI are inhibitors of CYP2C19 and might therefore inhibit the (cardioprotective) effect of clopidogrel. It is currently unclear whether this *in vitro* observation translates into a clinically relevant drug-drug-interaction, because the clinical effect might have been confounded as patients with concomitant use of PPI and clopidogrel were older and had more cardiovascular risk factors.

Table 2. PPI as risk factors for nosocomial pneumonia and *Clostridium difficile*–associated diarrhea, data show univariate and multivariate regression analyses from 3286 medical ICU patients [5]

PPI as a risk factor	Nosocomial pneumonia (OR, 95%-CI)	*Clostridium difficile*–associated diarrhea (OR, 95%-CI)
Univariate	1.79 (1.47-2.18)	3.50 (1.87-6.55)
Multivariate	1.28 (0.95-1.73; n.s.) *	3.11 (1.11-8.74) **

* Adjusted for use of mechanical ventilation, age, sex, the need for renal replacement therapy, vasopressors, corticosteroids.

** Adjusted for use of different antibiotics, mechanical ventilation, corticosteroids, vasopressors, mechanical ventilation, malignancy and the need for renal replacement therapy.

Until this issue is unequivocally settled, we recommend separating the time of application, preferentially using pantoprazole (the PPI with the least interaction potential) or possibly switching from clopidogrel to ticagrelor, which lacks the need for enzymatic activation.

Other known side effects of PPI might be also relevant for ICU patients, such as hypomagnesaemia, which even let to a recent warning through the Food & Drug Administration [35], or the osteopenia associated with long-term administration of PPI and an increased rate of bone fractures [36]. It is unclear whether these issues have an impact on the prognosis of ICU patients receiving PPI for stress ulcer prophylaxis or for GI bleedings.

Non-Pharmacological Alternatives

Enteral nutrition might have a beneficial role in preventing stress ulcer-related complications at the ICU. A recent meta-analysis of 1836 patients uncovered that in patients receiving enteral feeding a pharmacological stress ulcer prophylaxis did not influence the risk for an episode of clinical important bleeding (OR 1.26; 95%-CI 0.43–3.7). Interestingly, in exactly this subgroup an acid suppressing strategy did increase the risk for nosocomial pneumonia (OR 2.81; 95%-CI 1.20–6.56). In patients receiving both an acid suppression and enteral feeding, the mortality was even significantly increased [19]. However, the latter results have not yet been confirmed in controlled trials designed to investigate this concrete hypothesis. Still, those factors stress once again the need to adapt the individual stress ulcer prophylaxis strategy to the individual risk.

Key Points

- Stress-related mucosal disease (SRMD) is common in critically ill patients and renders patients at the intensive care unit (ICU) at risk for upper gastrointestinal hemorrhages.
- PPI administered for stress ulcer prophylaxis can effectively decrease bleeding rates in patients with major risk factors and are superior to H2RA.
- ICU patients without risk factors should not receive prophylaxis.
- There is no evidence for a mortality benefit for patients receiving stress ulcer prophylaxis.
- Nosocomial pneumonia and *Clostridium difficile*–associated diarrhea are potential and serious complications of stress ulcer prophylaxis.
- The potential protective role of enteral feeding in stress ulcer prophylaxis requires further investigation.

References

[1] Lin P-C, Chang C-H, Hsu P-I, Tseng P-L, Huang Y-B. The efficacy and safety of proton pump inhibitors vs histamine-2 receptor antagonists for stress ulcer bleeding prophylaxis among critical care patients: A meta-analysis. *Crit. Care Med.* 2010;38(4):1197–205.

[2] Laine L, Takeuchi K, Tarnawski A. Gastric Mucosal Defense and Cytoprotection: Bench to Bedside. *Gastroenterology* 2008;135(1):41–60.

[3] Alhazzani W, Alenezi F, Jaeschke RZ, Moayyedi P, Cook DJ. Proton Pump Inhibitors Versus Histamine 2 Receptor Antagonists for Stress Ulcer Prophylaxis in Critically Ill Patients: A Systematic Review and Meta-Analysis*. *Crit. Care Med.* 2013;41(3):693–705.

[4] Alhazzani W, Alshahrani M, Moayyedi P, Jaeschke R. Stress ulcer prophylaxis in critically ill patients: review of the evidence. *Pol. Arch. Med. Wewn.* 2012;122(3):107–14.

[5] Buendgens L, Koch A, Bruensing J, et al. Administration of proton pump inhibitors in critically ill medical patients is associated with increased risk of developing Clostridium difficile-associated diarrhea. *Journal of Critical Care* 2014;29(4):696.e11–5.

[6] Bardou M, Quenot J-P, Barkun A. Stress-related mucosal disease in the critically ill patient. *Nature Reviews Gastroenterology & Hepatology* 2015;12(2):98–107.

[7] Kamada T, Sato N, Kawano S, Fusamoto H, Abe H. Gastric mucosal hemodynamics after thermal or head injury. A clinical application of reflectance spectrophotometry. *Gastroenterology* 1982;83(3):535–40.

[8] Laine L, Takeuchi K, Tarnawski A. Gastric Mucosal Defense and Cytoprotection: Bench to Bedside. *Gastroenterology* 2008;135(1):41–60.

[9] Björne H, Govoni M, Törnberg DC, Lundberg JO, Weitzberg E. Intragastric nitric oxide is abolished in intubated patients and restored by nitrite. *Crit. Care Med.* 2005;33(8):1722–7.

[10] Michida T, Kawano S, Masuda E, et al. Endothelin-1 in the gastric mucosa in stress ulcers of critically ill patients. *Am. J. Gastroenterol.* 1997;92(7):1177–81.

[11] Ritchie WP. Acute Gastric Mucosal Damage Induced by Bile Salts, Acid, and Ischemia. *Gastroenterology* 1975;68(4):699–707.

[12] Cook DJ, Fuller HD, Guyatt GH, et al. Risk factors for gastrointestinal bleeding in critically ill patients. Canadian Critical Care Trials Group. N. *Engl. J. Med.* 1994;330(6):377–81.

[13] Quenot J-P, Thiery N, Barbar S. When should stress ulcer prophylaxis be used in the ICU? *Curr. Opin. Crit. Care* 2009;15(2):139–43.

[14] Cook DJ, Griffith LE, Walter SD, et al. The attributable mortality and length of intensive care unit stay of clinically important gastrointestinal bleeding in critically ill patients. *Critical Care* 2001;5(6):368–75.

[15] Dellinger RP, Levy MM, Rhodes A, et al. Surviving Sepsis Campaign: international guidelines for management of severe sepsis and septic shock, 2012. *Intensive Care Med.* 2013;39(2):165–228.

[16] Reinhart K, Brunkhorst FM, Bone H-G, et al. [Prevention, diagnosis, treatment, and follow-up care of sepsis. First revision of the S2k Guidelines of the German Sepsis Society (DSG) and the German Interdisciplinary Association for Intensive and Emergency Care Medicine (DIVI)]. *Anaesthesist.* 2010;59(4):347–70.

[17] Kahn JM, Doctor JN, Rubenfeld GD. Stress ulcer prophylaxis in mechanically ventilated patients: integrating evidence and judgment using a decision analysis. *Intensive Care Med.* 2006;32(8):1151–8.

[18] Cook DJ, Reeve BK, Guyatt GH, et al. Stress Ulcer Prophylaxis in Critically Ill Patients: Resolving Discordant Meta-analyses. *JAMA* 1996;275(4):308–14.

[19] Marik PE, Vasu T, Hirani A, Pachinburavan M. Stress ulcer prophylaxis in the new millennium: a systematic review and meta-analysis. *Crit. Care Med* 2010; 38(11):2222-8.

[20] Messori A, Trippoli S, Vaiani M, Gorini M, Corrado A. Bleeding and pneumonia in intensive care patients given ranitidine and sucralfate for prevention of stress ulcer: meta-analysis of randomised controlled trials. *BMJ* 2000;321(7269):1103–3.

[21] Heidelbaugh JJ, Inadomi JM. Magnitude and economic impact of inappropriate use of stress ulcer prophylaxis in non-ICU hospitalized patients. *Am. J. Gastroenterol.* 2006;101(10):2200–5.

[22] Cook D, Guyatt G, Marshall J, et al. A Comparison of Sucralfate and Ranitidine for the Prevention of Upper Gastrointestinal Bleeding in Patients Requiring Mechanical Ventilation. *N. Engl. J. Med.* 1998;338(12):791–7.

[23] Pongprasobchai S, Kridkratoke S, Nopmaneejumruslers C. Proton pump inhibitors for the prevention of stress-related mucosal disease in critically-ill patients: a meta-analysis. *J. Med. Assoc. Thai.* 2009;92(5):632–7.

[24] Barkun AN, Bardou M, Pham CQD, Martel M. Proton pump inhibitors vs. histamine 2 receptor antagonists for stress-related mucosal bleeding prophylaxis in critically ill patients: a meta-analysis. A*m. J. Gastroenterol.* 2012;107(4):507–20–quiz521.

[25] Laine L, Jensen DM. Management of patients with ulcer bleeding. *Am. J. Gastroenterol.* 2012;107(3):345–60–quiz361.

[26] Sachar H, Vaidya K, Laine L. Intermittent vs Continuous Proton Pump Inhibitor Therapy for High-Risk Bleeding Ulcers: A Systematic Review and Meta-analysis. *JAMA Intern. Med.* 2014;174(11):1755–62.

[27] Thorens J, Froehlich F, Schwizer W, et al. Bacterial overgrowth during treatment with omeprazole compared with cimetidine: a prospective randomised double blind study. *Gut* 1996;39(1):54–9.

[28] Agastya G, West BC, Callahan JM. Omeprazole inhibits phagocytosis and acidification of phagolysosomes of normal human neutrophils in vitro. *Immunopharmacol & Immunotoxicol* 2000;22(2):357–72.

[29] Zedtwitz-Liebenstein K, Wenisch C, Patruta S, Parschalk B, Daxböck F, Graninger W. Omeprazole treatment diminishes intra- and extracellular neutrophil reactive oxygen production and bactericidal activity. *Crit. Care Med.* 2002;30(5):1118–22.

[30] Herzig SJ, Vaughn BP, Howell MD, Ngo LH, Marcantonio ER. Acid-Suppressive Medication Use and the Risk for Nosocomial Gastrointestinal Tract Bleeding. *Arch. Intern. Med.* 2011;171(11).

[31] Laheij RJF, Sturkenboom MCJM, Hassing R-J, Dieleman J, Stricker BHC, Jansen JBMJ. Risk of community-acquired pneumonia and use of gastric acid-suppressive drugs. *JAMA* 2004;292(16):1955–60.

[32] Johnstone J, Nerenberg K, Loeb M. Meta-analysis: proton pump inhibitor use and the risk of community-acquired pneumonia. *Aliment Pharmacol. Ther.* 2010;31(11):1165–77.

[33] Aseeri M, Schroeder T, Kramer J, Zackula R. Gastric Acid Suppression by Proton Pump Inhibitors as a Risk Factor for Clostridium Difficile-Associated Diarrhea in Hospitalized Patients. *Am. J. Gastroenterol.* 2008;103(9):2308–13.

[34] Ho PM, Maddox TM, Wang L, et al. Risk of Adverse Outcomes Associated With Concomitant Use of Clopidogrel and Proton Pump Inhibitors Following Acute Coronary Syndrome. *JAMA* 2009;301(9):937–44.

[35] Tamura T, Sakaeda T, Kadoyama K, Okuno Y. Omeprazole- and esomeprazole-associated hypomagnesaemia: data mining of the public version of the FDA Adverse Event Reporting System. *Int. J. Med. Sci.* 2012;9(5):322–6.

[36] Ito T, Jensen RT. Association of Long-Term Proton Pump Inhibitor Therapy with Bone Fractures and Effects on Absorption of Calcium, Vitamin B12, Iron, and Magnesium. *Curr. Gastroenterol. Rep.* 2010;12(6):448–57.

In: Proton Pump Inhibitors (PPIs)
Editor: Barbara Parker
ISBN: 978-1-63482-890-1

Chapter III

Use and Misuse of Proton Pump Inhibitors: A Survey on a General Population

L. Lombardo*[*], *A. Giostra*[1], *L. Viganò*[2], *M. Tabone*, *A. Grassi*[3], *M. Gionco*[4], *G. Rovera*[5], *A. Bo*[6], *P. Malvasi*[7], *and Appropriateness Study Mauriziano Hospital Group

Gastroenterology Dpt, Mauriziano U.Ist Hospital (1980-2012), Turin, Italy
Since 2012: Gastroenterology Service,
Poliambulatorio Statuto, Turin, Italy
[1]Health Physics Dpt, Mauriziano U.Ist Hospital, Turin, Italy.
[2]General Surgery Dpt, Mauriziano U.Ist Hospital, Turin, Italy.
[3]Endocrinology Dpt, Mauriziano U.Ist Hospital, Turin, Italy.
[4]Neurology Dpt, Mauriziano U.Ist Hospital, Turin, Italy.
[5]Reumatology Dpt, Mauriziano U.Ist Hospital, Turin, Italy
[6]Information Elaboration Center, Mauriziano U.Ist Hospital, Turin, Italy
[7]Hospital Management Direction, Mauriziano U.Ist Hospital, Turin, Italy

Source of support: NONE
Conflict of interest: NONE

[*] Correspondence Address: Lombardo Lucio, Gastroenterology Service, Poliambulatorio Statuto, Piazza Statuto 3-Via Manzoni 0, 10100-Turin, Italy. Tel 0039 11 548944. Cell 0039 333 1339417. e-mail:lombodilucio@yahoo.it

Abstract

Background: Prescription rates of Proton pump inhibitors (PPI) continue to rise with growing concern over possible side effects and costs.

Aim: To evaluate in real world: 1) the overall prevalence of short-term and long-term PPI use, 2) its relative appropriateness.

Materials & Methods: From January to April 2011, all the out-patients asking for medical or surgical consultation at the Mauriziano U.I Hospital in Turin were investigated with a structured computerized pro-forma for detailed pharmacological history (dosage, prescribing and time modalities). Patients taking PPI for at least 6 months over the last 2 years were defined "long-term users" (LTU). Diagnostic procedures, final diagnosis, symptoms and drugs were registered.

Results: Out of 1056 patients who entered the study (620 F, 436 M, median age 70 yrs, peak 75 yrs) 478 (306 F, 172 M; $p < 0.002$) were PPI users (45%). Among these, 373 were LTU (79%; F 238). Among LTU: 1) **Prescription** was made by the General Practitioner in 37%, by the Gastroenterologist in 30%, by other Specialists in 30% and as auto medication in 3% of the cases. 2) **Diagnostic procedures**: a previous UGIE was performed in 40%, oesophageal 24-h pHmetry in 2%, oesophageal manometry in 1% of the patients. Tests for Hp detection were performed in 45% of the patients. 3) **Diagnosis** was GERD in 31%, functional dyspepsia in 23%, unspecified gastritis in 7%, Hp-negative gastritis in 8%, Hp-positive gastritis in 10%, previous duodenal or gastric peptic ulcer in 5%, chronic atrophic gastritis in 4%, gastric neoplasia in 1%, diagnosis non specified in 11% of the cases. 4) **Symptoms**: when PPI treatment was discontinued pain/discomfort relapsed in 32%, continued to be absent in 38%, was not specified in 30%; at reintroduction of PPI pain/discomfort disappeared in 26%, persisted in 22%, partially regressed in 16%, was not specified in 36% of the patients. 5) Gastrolesive drugs were taken by 48% of patients.

Conclusions: This study confirms that PPIs use in general population is extensive (45%) with a statistically significant prevalence in female patients ($p<0.002$), being of long-term type in the majority of the cases (79%). In LTU patients lack of appropriateness was observed for diagnostic procedures in 58% and for indications in 69%. It is worth noting that the diagnosis was "chronic atrophic gastritis" in 4% and not even specified in 11% of the cases. Clinicians should be aware of these realities in order to avoid side effects and unjustified social costs.

Keywords: PPI, appropriateness, therapy, epidemiology

Introduction

Proton pump inhibitors (PPIs) represent one of the most commonly prescribed classes of drugs both in hospital and primary care settings (1, 2, 3). Their prescription rates continue to rise with growing concern over possible side effects (4, 5) and costs (6). Although dyspeptic and reflux symptoms are highly prevalent according to epidemiological surveys (7, 8) it is unknown if the increasing use of PPIs is related to a change in occurrence of the acid-related conditions or to a change in prescribing patterns. From the clinical point of view, there is a continuum between mild reflux symptoms occurring in non-erosive reflux disease (NERD) and severe gastro-oesophageal reflux disease (GORD). Both groups may at time access over-the-counter therapies (OCT). OCT treatment of typical reflux symptoms is now accepted as safe in short-term relief of symptoms (9). Other conditions currently recognised for treatment with PPIs are peptic ulcer, non-steroidal anti-inflammatory drugs and dyspepsia.

PPIs are often prescribed for minor symptoms over long-term period without a clear indication (2, 10). Knowledge regarding the type of patients using PPIs, why they use them and the consequences of this usage is limited.

Aim of this study is to evaluate in a general population referring to the Mauriziano Hospital in Turin, Italy: 1) the overall prevalence of short-term and long-term PPI use, 2) clinical and demographic characteristics of patients using PPIs, 3) its relative appropriateness.

Materials and Methods

From January to April 2011 all the out-patients seeking for medical or surgical consultation at the Mauriziano U.I Hospital in Turin, Piedmont, Italy, were submitted to thorough clinical examination and investigated through a structured computerized form for detailed pharmacological history with dosage, prescribing and time modalities. Patients taking PPIs for at least 9 months and those taking PPIs for at least 1 month and less than 9 months over the last 2 years were defined respectively as long-term users (LTU) and short-term users (STU). Diagnostic procedures, symptoms, diagnosis and drugs were registered. The study was approved by the local Ethical Committee (protocol n. 60506/C28.2, 2010). Prior to the beginning of the study two general meetings were held in our Hospital in order to obtain complete comprehension of the protocol, relative consent and homogenisation of clinical evaluations. Data were analysed by chi square test to evaluate difference among groups. A value of $p<0.05$ was considered statistically significant.

Results

Out of 1065 patients 1056 (99.15%) were eligible for the study and were statistically evaluated. Nine patients were excluded for lack of answer to one or more questions. Eight hundred seventy eight cases were recruited from the medical area and 178 from the surgical area. Out of 1056 eligible patients (620 females, 436 males) 478 were PPI users (45%), 306 being females and 172 males, with a statistically significant difference ($p<0.002$).

Among the PPI users 373 were LTU (79%) and 105 (21%) were STU ($p<0.001$).

Age and Gender

The age classes for each gender among PPI users are shown in figure 1.

All the age classes, from 20 to 80 years, resulted to be PPI users, with a significant increase in the classes 60-80 among LTU *vs* STU ($p<0.001$).

Males : 172
Females: 306

Age classes :years	10-20	21-30	31-40	41-50	51-60	61-70	71-80	81-90
Males	1	16	18	23	34	36	32	12
Females	10	35	34	42	54	79	34	18

Figure 1. Age classes and gender of 478 patients taking PPIs. Absolute numbers.

Drug Dosage and Time

Lansoprazole was taken by 35% of the patients, esomeprazole by 30%, omeprazole by 20%, pantoprazole by 10% and rabeprazole by 5% of the patients. The dosage was continuously full in 45% of the patients, full and then followed by a maintenance dosage in 35%, and continuously half in 20%, for a median period of time of 6.5±1, 4±0.5 and 4.5±2.5 years, respectively.

The most adopted treatment pattern was "uninterrupted", or with very short treatment-free intervals (60%), while "seasonal" pattern accounted for 25% and "à la démande" pattern for 15%.

Prescriptions

As far as the prescriptions are concerned, General Practitioners were responsible for 37%, Gastroenterologists for 35%, non-gastroenterological Specialists for 25% while an auto-medication regimen was registered in 3% of the cases. Out of the LTU patients, 40% were submitted to upper endoscopy, 2% to 24-hours pH-metry, 1% to oesophageal manometry and 1% to Rx digestive series.

Gastroenterological Diagnosis

The gastroenterological diagnosis was GORD in 31% of the cases, functional dyspepsia in 23%, *H.pylori*-positive Gastritis in 15%, *H.pylori*-negative gastritis in 10%, peptic ulcer in 5%, chronic atrophic gastritis in 4%, not specified gastritis in 11%, and gastric cancer in 1%. The female gender was significantly prevalent in the GORD group ($p<0.001$), in functional dyspepsia group ($p<0.001$) and in *H.pylori*-positive gastritis group ($p<0.001$), while no difference was registered in the other groups. The age range most prevalent was 60-80 years for the female patients in any group ($p<0.001$), (except gastric cancer and pancreatic neoplasia, probably because of paucity of the cases) while the male gender was uniformly scattered through all the age ranges.

H.pylori testing

A test to check *H.pylori* infection was performed by 45% of LTU patients (60% urea breath test, 22% serology, 18% histology).

Drugs

NSAIDs were taken by 24% of LTU, more frequently by female patients ($p<0.01$), aspirin by 12%, more frequently by male patients, although with no statistical significance; warfarin by 5%, ticlopidine by 5%, clopidogrel by 2%, and cytostatic medication by 6% without significant difference between males and females. A burden of 63% of LTU patients was consumers of multiple classes of drugs, mainly for cardiovascular conditions, hypertension, atherosclerosis, diabetes and thyroidopaties.

PPI Interruption

When PPI therapy was interrupted for any reason, pain and/or pyrosis relapsed in 32 % of LTU patients, while did not in 38%; lack of information was registered in 30% of patients. No difference was found between males and females. Once PPI therapy was re-introduced pain and/or pyrosis disappeared in 22%, lowered in intensity in 19%, persisted unchanged in 22%, while 37% of patients were unable to specify the effect.

Discussion

This study shows that a large proportion of the population (45%) seeking for medical/surgical advice as outpatients to a tertiary Hospital in Turin uses PPI drugs, with the majority of them in a long-term modality (79%). An extensive use of LTU of PPIs has been documented in many other Western countries such a UK, Denmark and Sweden (6), Ireland (11, 12), USA (13), and Australia (14) so that it seems to be a worldwide global phenomenon. Although the most frequent age classes are in the 60-80 years of age, a not negligible portion is found in the 30-50 years range (30%). This is a group for whom the appropriateness of prescribing PPIs is often questioned because of the everyday and non-life threatening nature of gastric disorders. Young patients (under 45 years) are a minority but, in absolute terms, a sizeable number who could potentially be taking PPIs for many years and therefore be inappropriately expensive (15). All types of PPIs seem to be involved, the most frequent being lansoprazole and omeprazole, with a continuous full dosage modality been adopted in nearly half of the cases (45%) for a median period of 6.5 years. Seasonal and "à la démande" modality of assumption, as a whole, was registered in 40% of the cases, while an uninterrupted modality was registered in 60% of the cases. The prescription was chiefly made by General Practitioners (37%) and by Gastroenterologists (35%). Although an auto medication was registered in a limited percent of the cases (3%), the absolute number of patients is not negligible and casts some concern on the appropriateness about the over-the-counter (OCT) PPI management. In fact, although OCT treatment of typical reflux symptoms is accepted as safe for short-term relief of symptoms (9) patients on long-term treatment require investigations, as do patients with alarm features. A large proportion of LTU patients (60%) failed an endoscopic diagnosis and only a minimal proportion of patients had diagnosis of GORD based on pH-metry (2%) or oesophageal manometry (1%). Although presumptive diagnosis of GORD could be made on symptoms evaluation basis, we think that the clinician should remind that the gold standard diagnostic test for GORD is the 24-hours pH-metry and that it should be used more frequently, especially in patients candidate to a long-term treatment or non-responsive to therapy, in order to avoid inappropriateness. A new simple test on the saliva, not invasive and repeatable, is currently available for a quick diagnosis of GORD and should probably deserve more attention by the clinicians (PEP-test, Biohit, HealthCare, Helsinki).

H.pylori testing, on the other hand, is probably used less frequently than it should be (45% of LTU). We think that test & treat policy could reduce inappropriateness of PPI long-term therapy. In this context a relatively new serological test, that provides functional information upon gastric function and *H. pylori* status, the GastroPanel (Biohit, HealthCare, Helsinki) can be of clinical interest (16). A significant proportion of patients (63% of LTU) were consumers of multiple classes of drugs, mainly for cardiovascular diseases, hypertension, atherosclerosis, diabetes and thyroidopaties, often for a generic, and incorrect, principle of "gastroprotection", with the obvious result of adding another drug to an already exuberant number of drugs. NSAIDs and aspirin were respectively assumed by 24% and 12 % of LTU patients. The rate of PPI treatment misuse is somewhat mirrored by the effects on symptoms following PPI treatment interruption: 38 % of the cases did not complain pain or pyrosis and 30% were not able to specify any difference between the treatment and interruption period. As a whole, they accounted for 68% of the cases. To the same conclusion leads the opposite observation: when PPIs were re-assumed pain persisted unchanged in 22% while the effect was not specified in 37% of the cases, altogether accounting for 59% of the cases.

In conclusion, this study confirms that PPI use, especially the long-term one, in general population is extensive and largely inappropriate with a significant prevalence in the female genre ($p<0.002$). In long-term PPI users lack of appropriateness was observed for diagnostic procedures in 58% and for indications in 69% of the cases. A large-scale specific education campaign in general population and General Practitioners seems to be urgently needed.

Acknowledgments

This study was carried out thanks to the work of all the clinicians of the Mauriziano Hospital in Turin. Special thanks to the *Appropriateness Study Mauriziano Hospital Group:*

Rigazio Caterina, Ercole Elena, Lavagna Alessandro, (Dpt Gastroenterology, Mauriziano Hospital, Turin).

Oliva Alessandro, Righini Paolo (Dpt Pneumology, Mauriziano Hospital, Turin)

Vitale Corrado, Michele Bruno, Bagnis Cristina (Dpt Nephrology, Mauriziano Hospital,Turin).

Capussotti Lorenzo, Amisano Marco, Vigano' Luca, Bertolino Franco, Mineccia Michela (Dpt General Surgery, Mauriziano Hospital, Turin)

Actis Dato Guglielmo, Brusca Renzo (Dpt Cardiovascular Surgery, Mauriziano Hospital, Turin)

Cumbo Pia (Dpt Vascular Surgery, Mauriziano Hospital, Turin)

Tondolo Enrico (Dpt Ear Nose and Throat Diseases, Mauriziano Hospital, Turin)

Trebini Franco, Maniscalco Michele (Dpt Neurology, Mauriziano Hospital, Turin)

References

[1] Gullotta R, Ferraris L, Cortelezzi C et al. Are we correctly using the inhibitors of gastric acid secretion and cytoprotective drugs? Results of a multicentre study. *It J Gastroenterol Hepatol.* 1997;29:325-9.

[2] Nardino RJ, Vender RJ, Herbert PN. Overuse of acid-suppressive therapy in hospitalized patients. *Am J Gastroenterol.* 2000; 95:3118-22.

[3] Naunton M, Peterson GM, Bleasel MD. Overuse of proton pump inhibitors. *J Clin Pharm Ther.* 2000; 25:333-40.

[4] Kwok CS, Loke YK. Meta-analysis: the effects of proton pump inhibitors on cardiovascular events mortality in patients receiving clopidogrel. *Aliment Pharmacol Ther* 2010; 31:810-23.

[5] Lombardo L, Foti M, Ruggia O et al. Increased incidence of small intestinal bacterial overgrowth during proton pump inhibitor therapy. *Clin Gastroenterol Hepatol* 2010;8:504-8.

[6] Touborg-Lassen. Acid related disorders and use of antisecretory medication. *Dan Med Bull.* 2007; 54:18-30.

[7] Guillemot F, Ducrotte P, Bueno L. Prevalence of functional gastrointestinal disorders in a population of subjects consulting for gastroesophageal reflux disease in general practice. *Gastroenterol Clin Biol.* 2005; 29:243-6.

[8] Ronkainen J, Aro P, Storkrubb T, Johansson SE et al. High prevalence of gastroesophageal reflux symptoms and esophagitis with or without symptoms in the general adult Swedish population: a Kalixanda study report. *Scand J Gastroenterol.* 2005; 40:275-85.

[9] Haag S, Andrews JM, Katelaris PH et al. Management of reflux symptoms with over the-counter proton pump inhibitors: issues and proposed guidelines. *Digestion.* 2009; 80:226-34.

[10] Strid H, Simren M, Bjorsson ES. Overuse of acid suppressant drugs in patients with chronic renal failure. *Nephrol Dial Transplant.* 2003; 18:570-5.

[11] Mat Saad AZ, Collins N, Lobo MM et al. Proton pump inhibitors: a survey of prescribing in an Irish general hospital. *Int J Clin Pract* 2005;59:31-4.

[12] Sebastian SS, Kernan N, Qasim A et al. Appropriateness of gastric antisecretory therapy in hospital practice. *Ir J Med Sci* 2003;172:115-7.

[13] George CJ, Korc B, Ross JS. Appropriate proton pump inhibitor use among older adults: a retrospective chart review. *Am J Geriatr Pharmacother* 2008;6:249-54.

[14] Huges JD, Tanpurekul W, Keen NC et al. Reducing the cost of proton pump inhibitors by adopting best practice. *Qual Prim Care* 2009;17:15-21.

[15] Grime JC, Pollock K. How do younger patients view long-term treatment with proto pump inhibitors? *J R Soc Promot Health* 2002;122:43-9.

[16] Lombardo L, Leto R, Molinaro GC et al. Prevalence of atrophic gastritis in dyspeptic patients in Piedmont. A survey using GastroPanel test. *Clin Chem Lab Med* 2010;48(9):1327-32.

In: Proton Pump Inhibitors (PPIs)
Editor: Barbara Parker

ISBN: 978-1-63482-890-1

Chapter IV

Reported Complications of Proton Pump Inhibitor Use: An Update for the Clinician

Edward C. Oldfield IV[1] *and*
David A. Johnson, MD, MACG, FASGE, FACP[2,*]
[1]Eastern Virginia Medical School, Norfolk, VA, US
[2]Department of Internal Medicine chief, Division of Gastroenterology, Eastern Virginia Medical School, Norfolk, VA, US

Abstract

Recently, proton pump inhibitors have been associated with a number of potential complications including impaired vitamin and mineral absorption, altered drug metabolism, and increased risk for infections. Subsequently, there has been a plethora of recent investigation into each of these areas, with conflicting results noted for the majority of the reported complications. This chapter aims to outline the basis for these reported risks, discuss the current clinical evidence, and also offer insight into the likely clinical significance of each complication.

* Corresponding author: David A. Johnson MD, MACG, FASGE, FACP. 885 Kempsville Road, Suite 114, Norfolk, VA 23502. E-mail: dajevms@aol.com.

I. Introduction

Proton pump inhibitors (PPIs) have become a mainstay of therapy not only in the practice of gastroenterology, but also for general medicine and even the general public. Having surpassed histamine-2 receptor antagonists (H2RAs) as the predominant acid suppressive medication, PPIs represent one of the highest grossing classes of pharmaceuticals with sales of $13.9 billion in 2010 [1]. Over the past 25 years of use, PPIs have been considered a generally safe class of medications, however, recent reports of potential complications associated with chronic PPI use have prompted a flurry of research and FDA product warnings. In many of these cases there is evidence to support a potential association, yet study designs have largely prohibited the establishment of any causal relationships. This chapter aims to outline the physiologic basis for each of the reported risks and then explore the current evidence-based literature, finally offering a clinical summary for each of the reported associations.

II. Effects on Vitamin and Mineral Absorption

Vitamin B12

The absorption of vitamin B12 is a multistep process dependent on a number of factors, one of which is gastric acid. After vitamin B12 is ingested, it must first be released from its dietary proteins so that it is able to bind to R-proteins, which will protect it from pancreatic enzymes. This release of B12 from dietary proteins is dependent on the activation of pepsin from its precursor, pepsinogen, by gastric acid. Once bound to R-proteins, the B12-R protein complex travels to the small intestine where it is broken down in the duodenum. Vitamin B12 is finally absorbed in the terminal ileum bound to intrinsic factor produced by the gastric parietal cells. Given that the absorptive pathway of B12 is partially dependent on gastric acid, long-term PPI therapy may lead to B12 deficiency.

A number of early clinical trials revealed conflicting evidence on the role of PPI therapy in B12 absorption and the development of B12 deficiency. Initial insight into this potential complication arose from a retrospective case-control study of the Idaho Medicaid database, which identified 125 case patients who started vitamin B12 supplementation during the study period [2].

These patients were then compared to 500 matched controls for rates of chronic acid suppression therapy, defined as PPI or H2RA use for >10 of the 12 months prior to first vitamin B12 injection. Overall, chronic acid suppression therapy was significantly associated with vitamin B12 deficiency (18.4% of case patients versus 11.0% of the control group; OR 1.82, 95% CI 1.08-3.09, p -0.025) [2]. An important limitation of this study, however, was that this patient population was largely female and of low socioeconomic status. As a result, the findings of this study may not be generalizable to the population as a whole. This is particularly true when considering that elderly patients appear to be at higher risk, with estimates placing the incidence of B12 deficiency near 20% in this population [3].

When examining the elderly population specifically, a similar finding was noted in a cross-sectional study of 542 patients over age 60, who were either residents of long-term care facilities or patients at an ambulatory geriatric clinic [4]. Among the 26% of patients who used PPIs, analysis showed that PPI use was associated with a decreased serum B12 of -5.4 pg/mL per month of PPI use ($p < 0.00005$, 95% CI -7.7 to -3.1); this effect was slowed, but not prevented by concomitant oral B12 supplementation ($p=0.0125$) [4]. A much smaller study among a group of 36 institutionalized patients age 60-89 years old, identified and treated long-term PPI users (>12 months) with 8 weeks of cyanocobalamin nasal spray and compared the rates of B12 deficiency in this group to 19 non-PPI users [5].

While the long-term PPI users were more likely to have a B12 deficiency, after receiving treatment they showed an increase in serum B12 concentrations ($p=0.012$) and also a decrease in the frequency of B12 deficiency ($p=0.004$), suggesting a beneficial role of B12 supplementation.

These finding conflicts with results from a study comparing vitamin B12 status in 125 couples over the age of 65 in which one partner was a long-term PPI users (>3 years) and the other was not [6]. After adjusting for age, gender, *Helicobacter pylori* status, C-reactive protein levels, and excluding couples in which either participant used nutritional supplements, there were no differences in mean vitamin B12 levels (345 pM vs. 339 pM, $p=0.73$) [6]. Additionally, there were no differences seen in the concentrations of the metabolic intermediates, methylmalonate or homocysteine; this is an important finding, as it is one of the only studies to examine these intermediates, which would likely be altered in a true B12 deficiency.

Most recently, a large case controlled study comparing the risk of acid suppressive medication in B12 deficiency was performed using the Kaiser Permanente Northern California database.

This study identified 25,956 patients with an incident diagnosis of vitamin B12 deficiency of whom 3120 (12.0%) received 2 or more years' supply of PPIs, 1087 (4.2%) received 2 or more years' supply of H2RAs, and 21,749 (83.8%) received neither PPI or H2RA. Both PPIs (OR 1.65, 95% CI 1.58-1.73) and H2RAs (1.25, 95% CI 1.17-1.34) were found to be significantly associated with an increased risk for vitamin B12 deficiency [7]. Additionally, higher frequency of PPI dosing (>1.5 pills per day) (OR 1.95, 95% CI 1.77-2.15) was more strongly associated with vitamin B12 deficiency than smaller doses (<0.75 pills per day) (OR 1.63, 95% CI 1.48-1.78) [7]. Importantly, the strength of the association diminished after discontinuation of therapy, suggesting that this is a reversible effect. When using an overall prevalence of 2.3% for individuals over 50 years of age and an OR of 1.65 predicted from this study, PPI use for >2 years yields a number need to harm (NNH) of 67 [7].

Clinical Summary

Gastric acid is an essential factor involved in the absorption of vitamin B12 from dietary proteins. As such, prolonged PPI use could lead to impaired absorption and subsequent B12 deficiency. The currently available literature offers conflicting results on this association, precluding any definitive conclusion. Further, to date no prospective trials have been conducted to establish causation. Despite this, the most recent evidence does support an increased risk for B12 deficiency among long-term PPI users and, in particular, the elderly who are a high-risk population. As a result, some authors suggest that periodic B12 monitoring of elderly patients on prolonged PPI therapy would be a prudent decision to identify and treat deficiency before complications develop [4]. Among patients who do develop B12 deficiency while on PPI therapy, B12 supplementation has been shown to increase serum B12 levels and decrease B12 deficiency [5].

Iron

Gastric acid plays an essential role in the absorption of dietary iron. The majority of dietary iron is consumed as nonheme iron (ferric, Fe^{3+}), which must be reduced to the ferrous (Fe^{2+}) form in order to be absorbed by the divalent metal transporter-1 in the duodenum [8]. This process is mediated not only by gastric acid, but also vitamin C, which additionally serves as a reducing and chelating agent to prevent the formation of insoluble compounds [9, 10].

As such, long-term therapy with PPIs may theoretically lead to impaired iron absorption through hypochlorhydric states or alterations in vitamin C activity. This has been evidenced by a study showing marked reductions in the concentration of vitamin C following 4 weeks of omeprazole therapy in healthy volunteers [11]; however, long term follow up of patients taking daily PPIs for up to 4 years has not shown a clinically significant impairment in iron absorption [12]. Even if iron absorption were to be impaired, the use of iron supplements absorbed independently of gastric acid and vitamin C can be used to treat this deficiency [13].

This finding was also shared with a recently published prospective trial of 9 healthy participants, showing that short term PPI use does not alter the absorption of orally administered iron supplementation [14].

Clinically, the concern for impaired iron absorption is an increased risk of developing iron deficiency anemia. This was the focus of a retrospective cohort study of 98 patients taking chronic PPI therapy (>1 year), which showed a significant decrease in hematologic indices from baseline: hemoglobin (-0.19 g/dL, $p = 0.03$), hematocrit (-0.63%, $p = 0.02$), and mean corpuscular volume (-0.49 fL, $p = 0.05$) [15]. Even after adjusting for confounders, chronic PPI use remained significantly associated with an increased risk for a decreased hemoglobin by 1.0 g/dL (OR 5.03, 95% CI 1.71-14.78, $p < 0.01$) and decreased hematocrit by 3% (OR 5.46, 95% CI 1.67-17.85, $p < 0.01$) [15].

While this study importantly highlights a potential association between chronic PPI use and anemia, there are several limitations that may preclude its application to the general population. First, it was a small retrospective study due to the exclusion of 1,042 of 1,140 patients meeting inclusion criteria. This suggests that in the average population taking chronic PPIs there are likely to be a number of factors that can potentially influence the risk for anemia as well, namely gastrointestinal conditions which place patients at higher risk for bleeding or malabsorption.

Lastly, even though a decrease in hemoglobin >1g/dL or hematocrit >3% is considered a clinically significant decrease, there was no way to evaluate if any of these patients were symptomatic and also no evaluation was performed to confirm the anemia was secondary to iron deficiency.

While there are certainly limitations to the available clinical trials concerning the risk for anemia and chronic PPI use, this association was also found in a recent review of cardiovascular patients taking PPIs. A retrospective review of 278 patients showed that anemia was much more frequent in PPI users versus non-users (51% and 19%, respectively).

Additionally, on multivariate analysis, independent risk factors for low hemoglobin included female sex, peripheral artery disease, low white blood cell count, old age, low glomerular filtration rate, and PPI use. When looking specifically at the 36 PPI users who had hemoglobin levels available from at least 1 year before and 1 year after PPI initiation, there was a significant decrease in hemoglobin of 0.38 g/dL (95% CI -0.67 to -0.09) [16]. It is important to note that at baseline patients taking a PPI tended to be older, have worse renal function, have higher rates of diuretic or antithrombotic therapy, and have higher rates of comorbidities. When combined with the multiple independent risk factors for low hemoglobin, there remains a potential for confounding that may not be explained in this study.

Clinical Summary

The absorption of dietary iron is dependent on gastric acid and also vitamin C. As such, chronic PPI use may impair iron absorption and lead to the development of iron deficiency anemia. To date, no prospective trials have been conducted to establish a causal effect. Two retrospective studies support an association between chronic PPI therapy and a decreased level of hemoglobin [15, 16]; however, given the large number of exclusion criteria in these studies and the potential for confounding from other risk factors, these results should be taken with caution until more clinical evidence is available. Lastly, in patients with iron deficiency, available literature suggests that PPIs do not interfere with the absorption of iron supplements that act independent of gastric acid and vitamin C.

Hypomagnesemia

In 2011 the FDA issued a warning statement concerning the potential for low magnesium levels based on a number of case reports among long-term PPI users [17-20]. While the current mechanism for this increased risk is unknown, it may be related to changes in intestinal absorption that occur with long term PPI use. Magnesium uptake within the intestine is mediated by both active and passive reabsorption; active transport is regulated by the transient receptor-potential melastatin 6 and 7 (TRPM 6/7) channels, while passive reabsorption is concentration dependent and regulated by tight junction complexes [21, 22].

In theory, PPIs may alter the luminal pH, resulting in a reduced extrusion of protons into the intestine and leading to decreased activity of the TRPM 6/7 channels [23].

Currently, the available clinical evidence offers conflicting data on both the role and clinical significance of long-term PPI use in the development of hypomagnesemia.

An initial systematic review identified 36 cases of PPI-induced hypomagnesemia (PPIH) noting a median occurrence after 5.5 years in these patients, while the onset ranged from as little as 14 days to 13 years [24]. An important finding in this study was that discontinuation of PPI led to resolution of the PPIH, however, re-challenge with PPI led to recurrence within 4 days; this recurrence was not seen when patients were treated with H2RAs. An important conclusion from this systematic review is that there is certainly a potential for PPIH with long-term PPI use; however, this is reversible with cessation of PPI therapy and does not recur with a change in therapy to H2RA.

More recently, there have been a number of studies closely reviewing the risk of hypomagnesemia with PPI use across a number of settings. To look at the frequency of hypomagnesemia in low-risk patients, a case-control study (154 patients, 84 controls) was performed in outpatients at a gastroenterology clinic who were on PPI therapy for at least 6 months and without chronic kidney disease or diuretic use [25]. Results of this study showed that magnesium levels did not significantly differ between PPI users and nonusers (2.17 ± 0.20 mg/dL and 2.19 ± 0.15, respectively) [25].

Specifically among high-risk patients, a cross-sectional study was performed on 512 consecutive renal transplant recipients, 20% of whom were on a PPI. In multivariable analysis PPI use was not an independent predictor of hypomagnesemia; rather, factors associated with lower serum magnesium were tacrolimus, cyclosporine, sirolimus, absence of mycophenolate mofetil, lower levels of parathyroid hormone, and higher eGFR [26]. The authors concluded that PPIs may be used in renal transplant recipients without particular concerning for PPIH. Together, these studies highlight the low prevalence of PPIH in both low and high risk populations, suggesting that PPIH may be more of a rare, idiosyncratic effect.

In contrast, the most recent studies have focused on the risk of PPIH for hospitalization and prolonged hospital stay. A higher prevalence of PPI use among patients with hypomagnesemia was noted in a cross section study of 5118 patients presenting to a large tertiary care emergency department.

In multivariable analysis, PPI use was significantly associated with hypomagnesemia (OR 2.1, 95% CI 1.54-2.85) even when adjusting specifically for pre-hospital diuretic use (OR 1.7, 95% CI 1.27-1.97) [27]. Additionally, PPI use was significantly associated with a longer hospitalization after adjusting for comorbidities (p <0.0001) [27].

Similar findings were noted in a population based case-controlled study looking at hospitalizations with hypomagnesemia in patients over 66 years of age, which showed that PPI use was associated with an overall adjusted OR of 1.43 (95% CI, 0.81-1.91) [28]. This risk was shown only to occur in patients with PPI use within the past 90 days, but not for those with recent (91-180 days) or remote (181-365 days) use. In subgroup analysis, PPI use was additionally associated with an increased risk for hypomagnesemia in patients also receiving diuretics (aOR 1.73, 95% CI 1.11-2.70) but not for those patients on PPI but not on diuretics (aOR 1.25, 95% CI 0.81-1.91) [28]. Notably, this finding was shared in a study observing the risk of hypomagnesemia and PPI use in 11,490 consecutive patients admitted to the intensive care unit at a tertiary care center. Among patients on receiving diuretic therapy, PPI use was significantly associated with an increased risk of hypomagnesemia (OR 1.54, 95% CI 1.22-1.95, p <0.001); however, there was no significant association for patients on PPI therapy but not diuretic therapy (OR 0.92, 95% CI 0.78-1.09, p = 0.35) [29]. In subgroup analysis, all diuretic classes were associated with an increased risk with PPI use, although this was most significant for loop diuretics [29]. Taken together, these studies suggest that clinicians should be aware of a potentially increased risk for hypomagnesemia in patients taking both PPIs and diuretics. While the renal handling of magnesium is normally preserved in PPIH [30], there may be an unknown mechanism through which PPI and diuretic interact, leading to alteration of this process. It is also possible that this involves changes to TRPM 6/7 channel functions, as genetic mutations of these channels have been shown to lead to hypomagnesemia [28, 29]. Further research may help to elucidate the role of these channels and their potential involvement, if any, in PPIH.

Clinical Summary

To date there have been numerous case reports and now multiple clinical studies showing an association between long term PPI use and the risk of hypomagnesemia. The most recent systematic review and meta-analysis (9 studies, 115,455 patients) also supports this association (OR 1.775, 95% CI 1.077-2.924), however, they also noted no definitive conclusion could be reached due to significant heterogeneity between studies ($i^2 = 98\%$, p <0.001) [31]. Given the differences in populations used in the numerous studies on this topic, this is not a surprising finding. Overall, clinicians should be aware for this potential interaction. In particular, this association appears to be most pronounced in patients on PPIs and concurrent diuretic therapy [28, 29].

When looking at the general outpatient population, the risk appears to be much smaller [25]. As such, close following of magnesium levels may be warranted in those higher risk patients, however, routine monitoring of magnesium in patients taking a PPI is likely unnecessary given the current state of evidence and lack of prospective randomized trials. Still, clinicians should be judicious in their use of PPIs and keep them were clinically indicated.

Calcium Metabolism and Bone Resorption

Among the more pressing concerns for prolonged PPI use is the potential for an increased risk of osteoporosis and osteoporosis-related fractures, which carry with them significant morbidity and mortality. In 2010 the FDA released a safety announcement concerning the potential for an increased risk of fractures among PPI users [32]. While this statement was later revised to limit the risk for PPI use longer than a short-term 2-week treatment course, it has since garnered much attention for this potential interaction in the medical literature. The pathophysiological basis of this association is multifactorial. First, gastric acid plays an essential role in the absorption of dietary calcium, which is consumed in the form of insoluble calcium salts and dependent on the acidic environment of the GI tract to release soluble ionized calcium [33, 34]. Therefore, hypochlorhydric states induced by PPIs could lead to impaired calcium absorption. Secondly, this impaired absorption of calcium may lead to a compensatory secondary hyperparathyroidism, thereby increasing the rates of osteoclastic bone resorption [35]. Lastly, PPIs have been shown to directly inhibit osteoclast activity by inhibiting the vacuolar H+/K+ ATPase (proton pump) [36]. Overall, the regulation of calcium metabolism and its relation to bone homeostasis is multifactorial and complexly regulated. This section will seek to address the physiologic evidence concerning the purported role of PPIs increasing the risk of osteoporosis and osteoporosis related factures. Subsequently, the currently available clinical studies will be analyzed and discussed before offering a clinical summary.

Physiologic Studies

While early studies suggested calcium absorption was impaired by PPIs [37-40], two recent well-conducted trials found no decrease in calcium absorption in either healthy young adults [41] or postmenopausal women [42] taking PPIs; rather, the studies suggest that only when calcium carbonate is given in the fasting state is there a negative effect on calcium absorption in the setting of reduced acid secretion. Importantly, 90% of intestinal calcium absorption is believed to occur in the small intestine. As a result, even in the absence of PPI therapy, the contents of the small intestine would be much more alkaline (pH 6.0-7.0) due to the other digestive secretions and less dependent on the acidic pH from gastric acid [42].

The impact of impaired gastric acidification, however, extends beyond just that of calcium absorption. Evidence from animal models has shown that genetic mutations in the vacuolar proton pump can lead to impaired calcium homeostasis; in order to prevent severe hypocalcaemia in this setting, there is a compensatory secondary hyperparathyroidism along with an increased number of osteoclasts, which functions to release calcium from mineralized bone and potentially leads to osteoporosis [43]. Importantly, the authors note that they were able to reverse these metabolic derangements through the administration of calcium gluconate, which along with calcium citrate is soluble at a neutral pH [43]. As such, they highlight the potential for these two calcium formulations, rather than calcium carbonate, to be used in the treatment of osteoporosis in patients with PPI-induced hypochlorhydria.

Another mechanism for this interaction may be impaired osteoclast activity secondary to PPI inhibition of the vacuolar proton pumps, which are necessary to create the acidic environment for bone resorption [44]. Evidence suggests that patients on PPIs have decreased osteoclast activity and bone resorption, as evidence by decreased levels of urinary calcium and hydroxyproline [45]. In additions, there appears to be evidence of new bone formation, marked by increased levels of the osteoblast precursors, osteocalcin and tissue-resistance alkaline phosphatase [45]. Despite the evidence of decreased bone resorption, there is the possibility that this disruption in bone homeostasis predisposes to osteoporosis by prohibiting normal bone turnover [46]. Further research on this topic is needed before any conclusions can be made.

Clinical Studies

One of the strongest arguments for an association between PPI use and risk of fracture came from a prospective cohort study of 79,899 postmenopausal women enrolled in the US Nurses Health study, which found an increased risk of hip fracture in women regularly using PPIs for at least two years compared to those not taking PPIs (HR 1.35, 95% CI 1.13-1.62) [47]. Additionally, longer PPI use was associated with increasing risk (P_{trend} <0.01) [47]. This association remained significant after adjusting for body mass index (BMI), physical activity, and calcium intake; however, when accounting for smoking history, PPI use was significant in current and former smokers (HR 1.51, 95% CI 1.20-1.91), but not for non-smokers (HR 1.06, 95% CI 0.77-1.46). This observation importantly highlights the potential for confounding when using observational data that may explain the positive association in this study. This potential for confounding was also noted in a systematic review and meta-analysis (10 studies, 223,210 fracture cases) in which PPI was found to be associated with an increased risk for hip fracture (OR 1.25, 95% CI 1.14-1.37) albeit with significant heterogeneity (P for heterogeneity = 0.002, i^2 = 65%); however, in subgroup analysis there was not only increasing heterogeneity (P <0.001, i^2 = 82%), but also no significant association between long-term PPI use and the risk of hip fracture (OR 1.30, (95% CI 0.98-1.70) [48]. The authors importantly note that given the observation nature of the studies used, any causal effect cannot be established due to the potential for unmeasured or residual confounding.

Across different populations, there have been a number of studies using different national health databases to assess a potential interaction between PPI use and fracture risk. Early evidence from the Canadian Multicentre Osteoporosis Study (CaMos) suggested the risk between PPI use and fracture risk may be more an observational effect than causal. In this population based study, patient monitoring of bone mineral density (BMD) was performed at baseline (8,340 patients), 5 years (6,458 patients), and 10 years (4,512 patients) to establish risk factors for osteoporosis and fracture risk [49]. At baseline, PPI users had a lower BMD compared to non-users; however, PPI use did not result in a significant decrease of BMD at 5 or 10 years, which remained non-significant when adjusting for potential confounders [49]. In contrast, a second study using this same database found that PPI use was associated with a shorter time to first non-traumatic fracture (HR 1.75, 95% CI 1.41-2.17, p<0.001) [50]. Baseline characteristics of these PPI users showed they tended to be more likely to have had a prior fragility fracture, to have used corticosteroids, or be on a bisphosphonate. These and additional risk factors, however, were adjusted for in a multivariate analysis finding a lower adjusted HR for PPI use of 1.40 (95% CI 1.11-1.77, p=0.004) [50]. In contrast, there was no significant association noted using an adjusted HR for time-dependent PPI exposure when examining time to first hip fracture (HR 1.75, 95% CI 0.94-3.26, p = 0.79) [50]. Combined these two studies show that PPI users have a higher level of baseline risk factors for facture, however, even when controlling for these risks, PPI use is still associated with an increased risk for fractures that is not dependent on the duration of PPI therapy.

A second population based study using the Korean Health Insurance Review and Assessment Service database has shared a similar association between PPI use and hip fractures (aOR 1.34, 95% CI 1.24-1.44) [51].

In addition, subgroup analysis revealed that PPI and bisphosphonate use showed a significant statistical interaction (p <0.01); current PPI users or high cumulative dose PPI users both had an increased risk of hip fractures, which decreased as PPI use became less recent or cumulative dose of PPI decreased (p trend <0.001) [51].

One hypothesis for this interaction is the potential for variation in compliance between PPI users and non-users, however no significant difference in complicates rates for bisphosphonate use in PPI users versus non-users was found when using the medication possession ratio (p = 0.09). The authors suggest that this relationship between PPIs and bisphosphonates may explain some of the previously inconsistent results in other studies that were subject to significant heterogeneity.

Using the United Kingdom primary care research database, a cohort study with nested case-control analysis was performed, looking closely at the risk of PPI dose and duration on the risk of fracture [52]. Overall, there was a slight increase in the adjusted OR for risk of fracture in PPI users (OR 1.09, 95% CI 1.01-1.17). Additionally, higher doses of PPI were associated with an increased risk for fractures (OR 1.31, 95%CI 1.06-1.61); however, there was no associated duration effect with fracture risk in patients with less than one month (OR 1.16, 95% CI 0.94-1.43) or greater than 5 years (OR 1.02, 95% CI 0.87-1.20) of PPI use. Interestingly, high dose H2RA use was associated with an OR of 2.77 (95% CI 1.21-6.37).

This is an important point as a strong basis for the increased risk of PPI use if the inhibition of the osteoclast vacuolar proton pump, which would not be affected by H2RAs. Therefore, this observation would suggest the increased fracture risk is more closely related to acid suppressive therapy, regardless of the class, rather than a PPI effect on osteoclast activity.

Most recently, a parallel-randomized trial was conducted to assess the effect of PPI use on bone turnover in which 39 postmenopausal women between the ages of 55 and 85 were randomized to receive 8 weeks of either pantoprazole, a PPI that irreversibly inhibits the H+-K+-ATPase, or revaprazan, a reversible antagonist of the H+-K+-ATPase [53]. Twenty-six patients completed 8 weeks of treatment and the effects on bone metabolism were assessed by measuring baseline and post-treatment serum calcium and urine deoxypyridinoline (DPD); patients receiving PPI showed both increased serum calcium ($p = 0.046$) and urine DPD ($p = 0.046$) [53].

In multivariate analysis, age >60 years was found to be independently associated with both increased serum calcium and urine DPD. As such, the authors suggest that there may be some age related effect that makes this older population more vulnerable to PPI-induced alterations in bone metabolism. At this time, however, there remains no established causal effect.

Another emerging hypothesis, which may explain some of the confounding in early studies, is the role of histamine in driving potential bone loss. Patients on long term PPIs have increased levels of gastrin, which can lead to gastric enterochromaffin like cell hyperplasia and their increased histamine release [54]. This increased histamine can then interact with osteoclasts, which express histamine receptors; importantly, it is the histamine type 1 receptor which is most effective in regulation bone resorption in mature osteoclasts [55]. Early evidence supporting this hypothesis comes from a case controlled trial using the Danish National Prescription Database to assess for a potential interaction between PPIs and histamine type 1 receptor antagonists (H1RA) [56]. Two interesting observations were noted in this study: first, irrespective of PPI use, H1RA use was associated with a lower risk of hip fracture than non-use (aOR 0.86, 95% CI 0.79-0.93); second, overall fracture risk was lower in patients on a PPI also taking an H1RA than those on a PPI alone (aOR 0.92, 95% CI 0.87-0.98), however, this was not the case for hip fracture specifically (aOR 0.99, 95% CI 0.85-1.16) [56].

Overall, this new observation for a potential interaction between H1RAs and PPIs emphasizes the complexity of the numerous pathways affecting bone regulation and homeostasis. It also brings into question a potential explanation for confounding that may have been seen in early clinical studies.

Clinical Summary

There exist a number of pathophysiological explanations for a potential increased risk of fractures in PPI users. Namely, impaired absorption of calcium, secondary hyperparathyroidism, inhibition of osteoclast vacuolar proton pump, and also histamine mediated osteoclast activation. The clinical evidence, however, is not so clear.

A number of early studies suggest no significant association [41, 42], yet more recent population based studies have found a significant association even when controlling for other risk factors [47, 49-53]. Despite this, confounding remains a strong potential given the wide overlap between the various risk factors for osteoporosis and fracture risk within the demographics of PPI users. Given the lack of randomized control trials, no definitive causal effect can be established.

Clinicians should be aware of this potential risk among PPI users and ensure that patients, in particular older postmenopausal patients, have an appropriate indication for PPI use. Additionally, as many of these patients are on calcium supplementation, if any concern lingers about the potential for impaired calcium absorption due to decreased gastric acidity, consideration should be given to prescribing a calcium supplement soluble at a neutral pH, either calcium gluconate or calcium citrate.

III. Alteration of Pharmacodynamics: Clopidogrel

PPIs are a recommended treatment option in addition to dual anti-platelet therapy during the initial phase of acute coronary syndrome, especially in patients with a history of GI bleeding or peptic ulcer [57-59]. There is concern, however, that the use of PPIs may interfere with the metabolism and function of certain antiplatelet agents, in particular clopidogrel. In 2009 the FDA issued an advisory statement against the concomitant use of all PPIs with clopidogrel, which has since been revised to include only the potent CYP2C19 inhibitors, omeprazole and esomeprazole [60].

While there has been a considerable amount of research supporting this potential interaction, the data from clinical studies is controversial and in some cases conflicts with results from *ex vivo* studies on platelet function.

Therefore, in order to understand this complex relationship between PPIs and clopidogrel, it is important to review the pharmacokinetics and pharmacodynamics of clopidogrel, the role of genetics in altering this metabolic pathway, and the currently available clinical data.

Pharmacokinetics

Clopidogrel is an inactive prodrug, which must undergo two oxidative steps for bioactivation into its active metabolite, a process mediated by the cytochrome P450 pathway. The first of these two steps converts clopidogrel to 2-oxo-clopidogrel relying on contributions from CYP2B6, CYP2C9, and CYP2C19 (35.8, 19.4, and 44.9% respectively); 2-oxo-clopidogrel is then converted to active metabolite by CYP2B6, CYP2C9, CYP2C19, and CYP3A4 (32.9, 6.76, 20.6, and 39.8% respectively) [61].

Importantly, CYP2C19 has a significant contribution to both oxidative steps and CYP3A4 has a large contribution to this second step. Additionally, all PPIs are metabolized by CYP2C19 and CYP3A4 [62]. Given this shared reliance on CYP enzymes for metabolism, in particular CYP2C19, it has been hypothesized that competition for these enzymes may lead to reduced activation of clopidogrel and subsequent adverse clinical outcomes.

The mechanism of CYP inhibition by PPIs, however, appears more complex than originally thought. Earlier studies showed that lansoprazole was the most potent inhibitor of CYP2C19, with omeprazole and esomeprazole showing inhibition to a lesser extent [62]. Subsequently, however, *in vitro* studies showed that omeprazole and esomeprazole, not lansoprazole, acted as metabolism-dependent irreversible inhibitors of CYP2C19 [63]. This was also tested in a randomized, 2-peroid crossover study of 160 healthy volunteers treated with clopidogrel 75mg/day with or without PPI assessing both the concentration of clopidogrel active metabolite and platelet function [64]. All PPIs tested showed a decrease in the peak plasma concentration of the active clopidogrel metabolite and potency on maximal platelet aggregation and platelet response units; in order of decreasing potency, these PPIs were omeprazole, esomeprazole, lansoprazole, and dexlansoprazole [64]. Even pantoprazole, which should theoretically have the least inhibition of clopidogrel activity, was also shown to significantly alter platelet activity in a randomized trial of 105 patients following percutaneous coronary intervention, even when adjusting for CYP2C19*2 allele status [65].

Supporting evidence for an alteration in the activity of clopidogrel comes from a double-blind placebo controlled trial in which 124 patients undergoing coronary artery stent implantation received dual-antiplatelet therapy (aspirin 75 mg/day and clopidogrel, loading dose followed by 75mg/day) in addition to either omeprazole 20 mg/day or placebo [66]. Platelet activity, assessed using the platelet reactivity index (PRI), was comparable at baseline (83.2% vs 83.9%), however, patients treated with omeprazole showed a significant decreased in the PRI by day 7 (39.8% vs 51.4%, $p < 0.0001$) [66].

Together these studies highlight the potential for PPIs to interfere with clopidogrel metabolism and lead to impaired platelet function as a result; however, the most recent systematic review looking at the effect of PPIs on platelet function in patients receiving clopidogrel found that there was considerable heterogeneity in study designs, patient characteristics, laboratory measurements of platelet function, and drug exposure [67]. Ultimately, the authors concluded that there is no clear or consistent interaction between PPIs and clopidogrel in these studies.

Genetic Polymorphisms

Making the significance of the PPI-clopidogrel interaction on clinical outcomes even more complicated is the role of genetic polymorphisms in CYP enzymes and their alterations in metabolism. Two reduced function CYP2C19 alleles, CYP2C19 *2 and *3, have been suggested to heavily influence the clinical outcomes in clopidogrel treated patients; the variant allele CYP2C19*2, in particular, accounts for more than 90% of the cases of poor metabolism [68, 69]. Importantly, these alleles show interethnic variation and are quite prevalent in certain populations, affecting nearly 30% of Caucasians, 40% of blacks, and 55% of East Asians [69, 70].

Additionally, there is a low rate of homozygosity for these alleles (<3% of all patients, roughly 10% of CYP2C19*2 carriers) suggesting that even carriers (heterozygotes) may have an increased risk [69]. In order to assess the impact these polymorphisms may have on clinical outcomes, a systematic meta-analysis (23 studies, 11,959 patients) was performed in patients treated with clopidogrel showing an increased risk of major adverse cardiovascular events (MACE) in CYP2C19*2 carriers compared to noncarriers (OR 1.29, 95% CI 1.12-1.49, $p < 0.001$) [69].

The CYP2C19*2 carriers also had an increased mortality (OR 1.79, 95% 1.10-2.91, $p = 0.019$) and risk for stent thrombosis (OR 3.45, 95% CI 2.14-5.57, $p < 0.001$) independent of baseline cardiovascular risk. A notable limitation of these results, however, was considerable statistical heterogeneity for MACE and stent thrombosis [69]. When considering the overall frequency of the CYP2C19*2 allele in this study was 28% (n = 3418), this percent seems comparable to the estimated Caucasian prevalence but lower than the East Asian prevalence. A second meta-analysis (7 studies, 8043 patients) also found an increased risk of MACE (RR 1.96, 95% CI 1.14-3.37, $p = 0.02$) and stent thrombosis (RR 3.82, 95% CI 2.23-6.54, $p = 0.0001$) among patients with CYP2C19*2 polymorphism [71].

Similarly, this meta-analysis also noted considerable heterogeneity and also evidence of publication bias on funnel plot. Overall, the results of these meta-analyses support an increased risk for MACE in CYP2C19*2 allele carriers; however, given the heterogeneity present among the included studies and the interethnic variation in CYP2C19*2 frequency, these results may not be presently applicable to current clinical practice until prospective trials addressing these variables have been performed.

Clinical Studies

There have been a number of studies exploring the risk of concomitant PPI and clopidogrel use as it pertains to clinical outcomes. Unfortunately, the results of these studies are conflicting, as many retrospective studies show an increased risk for adverse events [72-76], while others, including the only randomized control trial, show no increased risk [77-79].

Further complicating the picture, common limitations noted in nearly all studies include an increased number of comorbidities among PPI users, potential confounding, and the inability to account for all cardiovascular risk factors, compliance, or over-the-counter medication.

Table 1 outlines these major studies, including the study objective, design, patients, results, and limitations.

Since the publication of these studies there have been a number of systematic reviews and meta-analyses, which have yielded some interesting insight into this complicated interaction between clopidogrel and PPIs. One of the earlier meta-analyses (25 studies, 159,138 patients) found concomitant clopidogrel and PPI use led to an increased risk for MACE (RR 1.29, 95% CI 1.15-1.45, i^2 = 72%) and MI (RR 1.31, 95% CI 1.12-1.53, i^2 = 77%); however, it was not associated with an increased risk of mortality (RR 1.04, 95% CI 0.93-1.16, i^2 = 41%) nor a decreased the risk for the development of a gastrointestinal bleed (RR 0.50, 95% CI 0.37-0.69, i^2 = 28%) [80].

Importantly, subgroup analysis was performed to assess any difference in the risk for MACE between omeprazole and pantoprazole users, as pantoprazole has previously been suggested to have the least inhibition of clopidogrel activity of all the PPIs [63]. Results showed that neither omeprazole nor pantoprazole was associated with an increased risk for MACE, yet both analyses had notable heterogeneity [80].

Other important findings were significant overall heterogeneity and an increased number of comorbidities in PPI users, in particular diabetes mellitus and renal failure, which may themselves alter clopidogrel metabolism and also increase the prevalence of poly-pharmacy in these patients.

Table 1. Clinical studies evaluating the risk of concomitant clopidogrel and PPI use

Study	Objective	Design	Patients	Results	Comments	Limitations (noted by authors)
Ho et al. [72]	Assess all-cause mortality or rehospitalization in patients taking clopidogrel with or without PPI after hospitalization for ACS	Retrospective cohort of VA health records	8205 discharged on clopidogrel 5244 (63.9%) prescribed PPI 2961 (36.1%) no PPI	1) Death or rehospitalization: 615 w/out PPI (20.8%) vs 561 w/PPI (29.8%) 2) Multivariate analysis: PPI use on death or rehospitalization (aOR 1.25, 95% CI 1.11-1.41) 3) PPI use risk for: - recurrent ACS (aOR 1.86, 95%CI 1.57-2.20) - revascularization (aOR 1.49, 95% CI 1.30-1.71) - all cause mortality (aOR 0.91, 95% CI 0.80-1.05) 4) PPI use without clopidogrel aOR 0.98 (95% CI 0.85-1.13)	Patients with PPI were older and with more comorbidities PPI use: 59.7% omeprazole 2.9% rabeprazole 0.4% lansoprazole 0.2% pantoprazole 36.7% more than 1 type of PPI during follow-up No relationship between PPI dose and adverse outcomes (OR 1.00, 95% CI 0.99-1.01)	1) Prescription of PPI may be a marker of more severe comorbid conditions associated with adverse outcomes 2) Patients receiving PPIs may have more epigastric or atypical chest pain 3) No information available for recurrent ACS or revascularization outside of the VA health system
Juurlink et al. [73]	Assess the risk of co-administration of clopidogrel and PPI for reinfarction in patients following an acute MI	Population-based nested case-control study patients aged 66 years or older Data from Ontario Public Drug Program	13,636 patients 2,682 (19.7% prescribed PPI within 30 days) 4,224 (31.0% prescribed PPI within 90 days)	782 patients with reinfarction within 90 days Current PPI use associated with increased risk for reinfarction (aOR 1.27, 95% CI 1.03-1.57)	Pantoprazole was not associated with an increased risk for reinfarction (aOR 1.02, 95% CI 0.70-1.47) PPI other than pantoprazole (aOR 1.40, 95% CI 1.10-1.77)	1) no information on cardiac risk factors 2) No identification of over the counter medications, including ASA

Study	Objective	Design	Patients	Results	Comments	Limitations (noted by authors)
Rassen et al. [74]	Assess the risk of MI hospitalization, death, and revascularization among patients taking clopidogrel with or without PPI after hospitalization for ACS or PCI	Retrospective: 3 large cohorts of patients >65 years of age Locations: British Columbia - using the provincial health care system Pennsylvania New Jersey - using states' drug benefit programs data linked to Medicare Part A and B claims data	Total: 18,565 British Columbia: 10,391 1353 PPI use 9038 no PPI Pennsylvania: 4,176 1352 PPI use 2824 no PPI New Jersey: 3,998 1291 PPI use 2707 no PPI	1) MI hospitalization: 2.6% PPI vs 2.1% no PPI hd-PS aRR 1.22 (95% CI 0.99-1.51) 2) Mortality: 1.5% PPI vs 0.9% no PPI hd-PS aRR 1.20 (95% CI 0.84-1.70) 3) Revascularization 3.4% PPI vs. 3.1% no PPI hd-PS aRR 0.97 (95% CI 0.79-1.21)	Authors note: No observation of conclusive evidence for a clopidogrel-PPI interaction of major clinical relevance; if effect exists it is unlikely to exceed 20% risk increase Used more extensive confounding adjustment compared to previous studies, also verified this with sensitivity analysis	1) ASA use remained uncontrolled; although would be presumed to be taken by nearly all patients 2) US population may have been able to purchase PPIs over the counter 3) Revascularization up to discretion of physician – may explain some heterogeneity of the revascularization results
O'Donoghue et al. [77]	Assess the association between PPI use, measures of platelet function, and clinical outcomes for patients treated with clopidogrel or prasugrel	Post-hoc analysis of two randomized double-blind control trials	PRINCIPLE-TIMI 44: 201 total patients Clopidogrel (n=99) Prasugrel (n=102) PPI users (n=53)	Mean inhibition of platelet aggregation (PRINICPLE-TIMI 44): 1) significantly lower for PPI users than non-users 6 hr after 600mg clopidogrel loading dose (23.2% vs 35.2%, p=0.02)	CYP2C219 allele: TRITON-TIMI 38 trial: Random DNA samples taken from 1477 clopidogrel and 1466 prasugrel patients to test for CYP2C19 allele status	1) PPI use was not randomized in these trials 2) Post-hoc analysis 3) PPIs could be initiated or discontinued and there was no way to monitor compliance

Table 1. (Continued)

Study	Objective	Design	Patients	Results	Comments	Limitations (noted by authors)
O'Donoghue et al. [77]	Primary endpoint: cardiovascular death, myocardial infarction, or stroke	PRINICIPLE-TIMI 44 TRITON-TIMI 38 [159, 160]	TRITON-TIMI 38: 13608 total patients Clopidogrel (n=6795) Prasugrel (n=6813) PPI users (4529)	2) Lower non-significant difference between PPI and non PPI users following 60mg prasugrel loading dose prasugrel (69.6% vs 76.7%, p =0.054) Risk of cardiovascular event (TRITON-TIMI 38): No association between PPI use and risk of primary endpoint for either clopidogrel (HR 0.94, 95% CI 0.80-1.11) or prasugrel (HR 1.00, 95% CI 0.84-1.20)	Reduced function allele seen in 357 clopidogrel and 372 prasugrel patients; of these 34% of clopidogrel and 33% of prasugrel on PPI at time of randomization Among these patients in the clopidogrel group, primary endpoint occurred in 10.2% (12/120) of PPI users vs 13.0% (30/237) of non-PPI users (aHR 0.76, 95% CI 0.39-1.48) This risk was also non-significant in the prasugrel group; PPI use (aHR 0.81, 95% CI 0.39-1.85)	4) Subgroups may have been underpowered to show significant association 5) Genetic analysis was underpowered and should be viewed as exploratory
Stockl et al. [75]	Evaluate the adverse clinical outcomes (rehospitalization or revascularization) in patients taking clopidogrel with or without a PPI for 360 days following discharge from MI or PCI	Retrospective matched cohort of persons enrolled in multistate health insurance plan with commercial and Medicare clients	Before matching: 1041 with PPI 6008 no PPI	1) MI rehospitalization: increased risk for clopidogrel plus PPI (aHR 1.93, 95% CI 0.05-3.54, p = 0.03)	Pantoprazole subgroup: aHR 1.91 (95% CI 1.19-3.06, p = 0.008)	1) no evaluation of important cardiovascular risk factors: race/ ethnicity, family history, obesity, smoking status

Study	Objective	Design	Patients	Results	Comments	Limitations (noted by authors)
Stockl et al. [75]			After matching: 1033 for both PPI and non-PPI use	2) MI rehospitalization or revascularization: increased risk clopidogrel plus PPI (aHR 1.64, 95% CI 1.16-2.32, p = 0.005)	Clopidogrel plus PPI group: - higher mean Charlson comorbidity score (p=0.04)	2) no evaluation of cardiovascular deaths may underestimate the occurrence of MI 3) no control for ASA use 4) no comparison between type of stents used
van Boxel et al. [76]	Assess the risk and occurrence of cardiovascular and gastrointestinal events in co-administration of a PPI and clopidogrel Composite endpoint: MI, unstable angina, stroke, all cause mortality GI events: complicated or uncomplicated PUD	Retrospective of new clopidogrel users >18 years of age from two Dutch health insurance companies covering 4 million inhabitants (roughly 25% of the total population of the Netherlands)	18,139 new users of clopidogrel 5,7634 (32%) concomitant users of PPIs	In multivariate analysis, use of clopidogrel and PPI was associated with and increased risk of: 1) MI (HR 1.93, 95% CI 1.40-2.65) 2) Unstable angina (HR 1.79, 95% CI 1.60-2.03) 3) All cause mortality (HR 1.79, 95% CI 1.44-2.22) 4) Composite endpoint (HR 1.75, 95% CI 1.58-1.94) 5) GI events (HR 4.76, 95% CI 1.18-19.17)	PPI users were significantly older, used more co-medications, and suffered from more comorbidities	1) PPIs may be a marker for decreased health status and potentially increased CV risk 2) Risk for channeling bias and confounding by indication 3) Analysis exclusively from administrative claims data and outcomes are not validated by chart review

Table 1. (Continued)

Study	Objective	Design	Patients	Results	Comments	Limitations (noted by authors)
Bhatt et al. [78]	Assess the rates of adverse gastrointestinal and cardiovascular outcomes in patients taking clopidogrel and omeprazole versus clopidogrel alone	Randomized, double-blind, double-dummy, placebo-controlled, parallel-group, phase 3 study of the efficacy and safety of a fixed dose combination of clopidogrel (75mg) and omeprazole (20mg) compared to clopidogrel alone Median duration of follow-up 106 days	3,761 patients 1,876 omeprazole group 1,885 in placebo group	Gastrointestinal events (51 total): 1.1% omeprazole vs 2.9% placebo (HR 0.34, 95% CI 0.18-0.63, p <0.001) Rate of overt upper GI bleed reduced with omeprazole (HR 0.13, 95% CI 0.03-0.56, p =0.001) Cardiovascular events (109 total): 4.9% omeprazole vs 5.7% placebo (HR 0.99, 95% CI 0.68-1.44, p=0.96)	Rate of symptoms of GERD: 0.2% omeprazole vs 1.2% placebo (HR 0.22, 95% CI 0.06-0.79, p =0.01) Subgroup analysis did not show significant heterogeneity There was no significant difference in rate of adverse events (p=0.33)	1) Trial stopped early due to lack of funding 2) Absence of interaction between clopidogrel and omeprazole cannot be viewed as definitive as wide CI for HR in CV events 3) 94% of the study population was white – with expected loss-of-function CYP2C19 homozygosity of 2-3% 4) Single pill administration of clopidogrel and omeprazole has different release kinetics than generic omeprazole
Roe et al. [79]	Prasugrel versus Clopidogrel for acute coronary syndromes without revascularization	Double-blind, randomized trial assessing the efficacy of 30 months of treatment with prasugrel versus clopidogrel	7243 total patients 1666 PPI users	Event rate among PPI users: Prasugrel 14.6% vs Clopidogrel 23.8% (HR 0.70, 95% CI 0.53-0.92, p =0.02)		Subgroup analysis only

ACS: acute coronary syndrome, PCI: percutaneous coronary intervention, MI: myocardial infarction, hd-PS aRR: high-dimensional propensity score adjusted risk ratio, ASA: aspirin, PUD: peptic ulcer disease.

Table 2. Clinical Summaries

Reported Risk	Clinical Summary
Vitamin B12	Gastric acid is an essential factor involved in the absorption of vitamin B12 from dietary proteins. As such, prolonged PPI use could lead to impaired absorption and subsequent B12 deficiency. The currently available literature offers conflicting results on this association, precluding any definitive conclusion. Further, to date no prospective trials have been conducted to establish causation. Despite this, the most recent evidence does support an increased risk for B12 deficiency among long-term PPI users and, in particular, the elderly who are a high-risk population. As a result, some authors suggest that periodic B12 monitoring of elderly patients on prolonged PPI therapy would be a prudent decision to identify and treat deficiency before complications develop [4]. Among patients who do develop B12 deficiency while on PPI therapy, B12 supplementation has been shown to increase serum B12 levels and decrease B12 deficiency [5].
Iron	The absorption of dietary iron is dependent on gastric acid and also vitamin C. As such, chronic PPI use may impair iron absorption and lead to the development of iron deficiency anemia. To date, no prospective trials have been conducted to establish a causal effect. Two retrospective studies support an association between chronic PPI therapy and a decreased level of hemoglobin [15, 16]; however, given the large number of exclusion criteria in these studies and the potential for confounding from other risk factors, these results should be taken with caution until more clinical evidence is available. Lastly, in patients with iron deficiency, available literature suggests that PPIs do not interfere with the absorption of iron supplements that act independent of gastric acid and vitamin C.
Magnesium	To date there have been numerous case reports and now multiple clinical studies showing an association between long term PPI use and the risk of hypomagnesemia. The most recent systematic review and meta-analysis (9 studies, 115,455 patients) also supports this association (OR 1.775, 95% CI 1.077-2.924), however, they also noted no definitive conclusion could be reached due to significant heterogeneity between studies ($i^2 = 98\%$, $p < 0.001$) [31]. Given the differences in populations used in the numerous studies on this topic, this is not a surprising finding. Overall, clinicians should be aware for this potential interaction. In particular, this association appears to be most pronounced in patients on PPIs and concurrent diuretic therapy [28, 29]. When looking at the general outpatient population, the risk appears to be much smaller [25]. As such, close following of magnesium levels may be warranted in those higher risk patients, however, routine monitoring of magnesium in patients taking a PPI is likely unnecessary given the current state of evidence and lack of prospective randomized trials. Still, clinicians should be judicious in their use of PPIs and keep them were clinically indicated.

Table 2. (Continued)

Reported Risk	Clinical Summary
Calcium	There exist a number of pathophysiological explanations for a potential increased risk of fractures in PPI users. Namely, impaired absorption of calcium, secondary hyperparathyroidism, inhibition of osteoclast vacuolar proton pump, and also histamine mediated osteoclast activation. The clinical evidence, however, is not so clear. A number of early studies suggest no significant association [41, 42], yet more recent population based studies have found a significant association even when controlling for other risk factors [47, 49-53]. Despite this, confounding remains a strong potential given the wide overlap between the various risk factors for osteoporosis and fracture risk with the demographics of PPI users. Given the lack of randomized control trials, no definitive causal effect can be established. Clinicians should be aware of this potential risk among PPI users and ensure that patients, in particular older postmenopausal patients, have an appropriate indication for PPI use. Additionally, as many of these patients are on calcium supplementation, if any concern lingers about the potential for impaired calcium absorption due to decreased gastric acidity, consideration should be given to prescribing a calcium supplement soluble at a neutral pH, either calcium gluconate or calcium citrate.
Clopidogrel	There is a plethora of published material concerning the interaction between clopidogrel and PPIs. Early studies on pharmacokinetics showed PPIs altered clopidogrel activity on platelet function secondary to inhibition of CYP enzymes. This effect, originally thought to be primarily limited to the potent CYP2C19 inhibitors, was subsequently shown to be a class effect [63, 65]; however, recent analyses have suggested strong heterogeneity among these studies limits their generalizability [67, 69, 71]. Further, the clinical significance of this interaction is not consistent between clinical studies, and the multiple meta-analyses have all been subject to marked heterogeneity [80, 81, 83, 84]. These results question the validity of the observed increased risk for adverse clinical outcomes and suggest that at this time no conclusion can be reached. Additionally, it is possible that CYP2C19 polymorphisms are a major factor in predicting poor clinical outcomes, however, at this time no clear guidelines are available for recommended screening of these polymorphisms. Substitution of an H2RA for a PPI is likely of limited value in this setting, as H2RAs have still been shown to increase the risk for adverse cardiovascular outcomes, albeit a slightly lower risk than PPIs [84, 85]. Similarly, the use of other antiplatelet agents, in particular prasugrel, has been suggested as an alternative to clopidogrel, however, there is *ex vivo* evidence that PPIs may still interfere with the metabolism of these drugs [86]. In conclusion, equivocal outcomes in clinical studies preclude the recommendation for discontinuation of PPIs among patients on dual-antiplatelet therapy for whom there is an appropriate indication for acid suppressive therapy.

Reported Risk	Clinical Summary
Pneumonia	Numerous studies and meta-analyses have shown an apparent association between PPI use and an increased risk of pneumonia. In particular, this risk is most pronounced in the short-term setting with no significant association noted for long-term use [89-93]; however, the results of these studies should be interpreted with caution, as a significant level of heterogeneity precludes any definitive relationship [92-94]. Further, there is the possibility for a protopathic bias to explain this temporal association. Lastly, recent studies have suggested the potential for significant confounding in this relationship, as PPI use has been implausibly associated with common medical conditions in addition to PPI use [97]. As such clinicians should be aware of the potential short-term association of PPI use with pneumonia, however, continue to use PPI where appropriately indicated.
C. difficile infection	Overall, there is conflicting evidence to support a definitive relationship between PPI use and the development of CDI. Despite this, a number of meta-analyses have demonstrated a significant association even when attempting to control for risk factors [105-108]. These studies should be interpreted with some caution, however, as there was considerable heterogeneity and evidence of publication bias, constituting a very low-grade quality of evidence. Further, even with the positive association, the NNH for the general population is 3925 and for hospitalized patients not on antibiotics the NNH is 367 [108]. Lastly, neither the Infectious Disease Society of America (IDSA) nor the American College of Gastroenterology (ACG) recommend routine discontinuation of PPI use in their clinical guidelines for the diagnosis and treatment of CDI [112, 113]. As such, the benefits of PPI use in patients for whom there is an appropriate indication for acid suppression likely outweighs the risk of CDI, and PPI therapy should not be discontinued on the basis of a potential risk for CDI.
Small intestinal bacterial overgrowth	While there does appear to be a pathophysiological basis for an increased risk of SIBO in patients taking PPI, the current clinical evidence is controversial and prohibits any definitive conclusion. This is particularly true given the lack of randomized control trials, variability in diagnostic measures, and potentially unrecognized confounding. Still, clinicians should be aware of the potential risk for bowel symptoms and SIBO in patients receiving long-term PPI therapy. In patients who do develop SIBO, rifaximin 400mg three times daily for 14 days has been shown to effectively eradicate the majority of SIBO regardless of the duration of PPI therapy [118].
Spontaneous Bacterial Peritonitis	While initial studies suggested a minimal risk of PPI use associated with the development of SBP in cirrhotic patients with ascites, several recent studies have all supported an associated risk [126-128, 137, 138]. There are still certain limitations within these studies, namely many with small sample sizes and retrospective designs.

Table 2. (Continued)

Reported Risk	Clinical Summary
Spontaneous Bacterial Peritonitis	Additionally, there are a number of confounders in these cirrhotic patients that may not have been accounted for, in particular, when the majority of studies indicate the SBP patients tend to have more advanced liver disease. Despite this, it remains clear that clinicians should scrutinize the use of PPIs in this patient population and ensure that patients taking these medications have an appropriate indication for their use.
Traveler's Diarrhea	There is currently limited data to support a definitive risk for travelers' diarrhea among PPI users. While there is evidence to support an increased risk of enteric infections related to acid suppressive medications [104, 142], this has not been studied specifically in travelers. The International Society of Travel Medicine does, however, consider the use of acid suppressive medication to be a risk factor for enteric infections and recommends that patients taking these medications be considered for chemoprophylaxis when traveling to high-risk endemic areas [148].
Acute Interstitial Nephritis	Despite its rare and idiosyncratic occurrence, PPI-induced acute interstitial nephritis is a real clinical entity with the potential for severe sequelae. As such clinicians should have a high level of suspicion to detect AIN early in its presentation, particularly if patients were recently started on a PPI.
Methotrexate	There is evidence to suggest a potential mechanism whereby PPI co-administration with high-dose methotrexate may lead to delayed drug elimination and methotrexate toxicity. Clinical evidence from retrospective reviews, however, offers conflicting evidence on the prevalence and significance of this association. As such clinicians should be aware of this potential interaction and monitor patients closely for any signs of toxicity. Further, given that there are no reported interactions with H2RAs, physicians should consider switching therapy prior to high-dose methotrexate therapy.

This finding was shared with a systematic review (33 studies) which found significant heterogeneity in clinical outcomes (RR range 0.64-4.58) suggesting that imbalances between prognosticators at baseline and PPI prescription bias markedly contributed to the variability in the results [81].

The heterogeneity in these studies may be partially related to different risk profiles based on the study design.

Kwok et al. conducted a meta-analysis (23 studies, 93,278 patients) which found an increased risk for concomitant clopidogrel and PPI use among observational studies (aRR 1.54, 95% CI 1.23-1.92), however, propensity-matched or randomized control trials showed no association (RR 1.15, 95% CI 0.94-1.26) [82]. Two years later, these same authors conducted another meta-analysis (23 studies, 222,311 patients) attempting to improve the earlier study by including intra-class differences between PPIs [83].

Despite moderate-substantial heterogeneity, the individual PPIs, omeprazole, esomeprazole, lansoprazole, and pantoprazole, all showed increased risk for cardiovascular events when prescribed with clopidogrel.

Repeat analysis again showed an increased risk for MACE was seen in observational studies (OR 1.28, 95% CI 1.14-1.44), yet the two randomized controlled trials did not show any increased risk for omeprazole or esomeprazole [83].

Recognizing these variations in observed outcomes may be dependent on study design, a group of investigators conducted an observational cohort study and a self-controlled case series on 24,471 patients receiving dual-antiplatelet therapy to assess the impact of PPI use on death or incident MI [84]. In multivariable analysis, PPI use was associated with an increased risk for death or incident MI (HR 1.37, 95% CI 1.27-1.48).

Conversely, in the self-controlled case series there was no association between PPI use and MI (HR 0.75, 95% CI 0.55-1.01). This finding suggests that confounding between people may account for a large percentage of the attributed risk.

Further, all cause mortality was elevated not only among users of the CYP2C19 non-inhibiting drugs, ranitidine (aHR 1.20, 95% CI 0.99-1.46) and citalopram (aHR 1.52, 95% CI 1.32-1.76), but also the CYP2C19 inhibiting drugs, paroxetine and fluoxetine (aHR 1.42, 95% CI 1.17-1.72); these findings, which would not be expected solely based on pharmacokinetics, suggest that patients taking any long-term medical therapy may be at a higher risk of harmful outcomes [84].

Clinical Summary

There is a plethora of published material concerning the interaction between clopidogrel and PPIs. Early studies on pharmacokinetics showed PPIs altered clopidogrel activity on platelet function secondary to inhibition of CYP enzymes.

This effect, originally thought to be primarily limited to the potent CYP2C19 inhibitors, was subsequently shown to be a class effect [63, 65]; however, recent analyses have suggested strong heterogeneity among these studies limits their generalizability [67, 69, 71].

Further, the clinical significance of this interaction is not consistent between clinical studies, and the multiple meta-analyses have all been subject to marked heterogeneity [80, 81, 83, 84]. These results question the validity of the observed increased risk for adverse clinical outcomes and suggest that at this time no conclusion can be reached. Additionally, it is possible that CYP2C19 polymorphisms are a major factor in predicting poor clinical outcomes, however, at this time no clear guidelines are available for recommended screening of these polymorphisms.

Substitution of an H2RA for a PPI is likely of limited value in this setting, as H2RAs have still been shown to increase the risk for adverse cardiovascular outcomes, albeit a slightly lower risk than PPIs [84, 85]. Similarly, the use of other antiplatelet agents, in particular prasugrel, has been suggested as an alternative to clopidogrel, however, there is *ex vivo* evidence that PPIs may still interfere with the metabolism of these drugs [86].

In conclusion, equivocal outcomes in clinical studies preclude the recommendation for discontinuation of PPIs among patients on dual-antiplatelet therapy for whom there is an appropriate indication for acid suppressive therapy.

IV. Effects on Risk for Infection

Pneumonia

Proton pump inhibitors have recently been linked to an increased risk of both community-acquired and hospital-acquired pneumonia. A theoretical basis for this increased risk of pneumonia comes from evidence suggesting that patients with gastroesophageal reflux disease (GERD) on PPI therapy have an increased prevalence of gastric bacterial overgrowth in comparison to both those taking an H2RA or not on therapy [87, 88]. This colonization may then lead to an increased risk for aspiration pneumonia.

Given the potential economic and health implications for an increased risk of pneumonia, correctly assessing the importance of this association has been of great interest.

Despite a number of studies showing a positive association, the available literature is plagued by heterogeneity and the potential for confounding.

Community Acquired Pneumonia

One of the initial reports suggesting an increased risk of community-acquired pneumonia (CAP) came from a population based case-control study in Denmark, reviewing 34,176 patients and revealing an adjusted OR of 1.5 (95% CI 1.3-1.7) for current PPI use in the development of CAP [89].

This association was not found for H2RAs (OR 1.1, 95% CI 0.8-1.3) or for previous PPI use (OR 1.2, 95% CI 0.9-1.6). In particular, the strongest association was found in the subgroup with PPI use 0 to 7 days prior to the index date (OR 5.0, 95% CI 2.1-11.7). These overall results, however, conflicted with a similar nested case-control study in the United Kingdom, which found that current PPI use was not associated with an increased risk for CAP (OR 1.02, 95% CI 0.97-1.08) [90].

Interestingly, this study shared the same temporal association for recent PPI use: PPI use within the previous 2 days (OR 6.53, 95% CI 3.95-10.80), 7 days (OR 3.79, 95% CI 2.66-5.42), 14 days (OR 3.21, 95% CI 2.46-4.18) [90]. Combined, these studies suggest that short-term use of PPIs (<7-14 days) may confer an increased risk of pneumonia.

More recently, a review of the New England Veterans Healthcare system comparing current with past PPI use also found an increased risk in patients currently taking a PPI (OR 1.29, 95% CI 1.15-1.45); the risk was increased in those with PPI exposures between 1 and 15 days as compared to those with longer exposures [91]. This study, however, may have been subject to confounding, as congestive heart failure, chronic obstructive pulmonary disease, lung cancer, chronic kidney disease, and those with >3 comorbidities were found to have a decreased risk of CAP [91]. The authors note that a potential explanation for this finding would be that these comorbidities may already be at high risk for CAP, thereby diminishing the additive risk of PPIs.

Two meta-analyses published on the topic share the same overall findings as these three studies, namely that recent, not long term, PPI use is associated with an increased risk of CAP. The first (6 studies, approximately 1 million patients) identified a significant OR of 1.92 (95% CI 1.40-2.63) for short duration of PPI use, compared to a non-significant OR of 1.11 (95% CI 0.90-1.38) for chronic PPI use [92]. Importantly, there was a considerable heterogeneity that precluded any interpretation of the results (overall i^2 = 92%). The second meta-analysis (9 studies, 120,863 patients) found significant associations with current PPI use (OR 1.39, 95% CI 1.09-1.76), PPI use <30 days (OR 1.65, 95% CI 1.25-2.19), high dose PPI use (OR 1.50, 95% CI 1.33-1.68), and low dose PPI use (OR 1.17, 95% CI 1.11-1.24); however, there was no association with PPI use >180 days (OR 1.10, 95% CI 1.00-1.21) [93].

Contrasting evidence comes from the most recent meta-analysis which did not find any association between increased risk for hospitalization from community-acquired pneumonia and PPI use [94]. Results showed that neither PPI use (OR 1.05, 95% CI 0.89-1.25) nor H2RA use (0.95, 95% CI 0.75-1.21) was significantly associated with an increased rate of hospitalization of CAP in this patient population [94]. A strength of this study was an attempt to minimize bias by including only new users of NSAIDs, who were also prescribed prophylactic PPIs; the use of this patient population helps to better control for the protopathic effect that may result in the positive temporal association seen in other studies. The notion of the protopathic bias, in which a pharmaceutical agent is inadvertently prescribed for an early manifestation of a disease that has not yet been diagnostically detected, represents an important aspect of the potential association between PPI use and CAP. This is of particular concern in the relationship between PPIs and pneumonia, as the strongest associations are noted within 7 days, even within 2 days [89, 90]. Given that PPIs may not have achieved their full effect in this timeframe, this may represent a protopathic bias; patients may be prescribed PPIs for early signs of pneumonia, such as nonspecific chest symptoms or discomfort [94].

Hospital Acquired

The risk of hospital-acquired pneumonia in patients receiving acid suppressive medication is a critical issue, as it is reported that nearly 50% of all in-patients are started on these medications for stress ulcer prophylaxis regardless of indication [95]. To assess the association between acid suppressive medication and hospital-acquired pneumonia, researchers conducted a prospective pharmacoepidemiologic cohort study of 63,878 admissions to a single academic medical center. Results showed that PPI use (OR 1.3, 95% CI 1.1-1.4), but not H2RA use (OR 1.2, 95% CI 0.98-1.4), was associated with an increased risk for pneumonia [96]. This highlights an important point, as using these study results (52% of patients on acid suppressive medication and an overall rate of hospital acquired pneumonia at 3.5%) there is an attributable risk of 0.9%; this translates in to an NNH of 111. This NNH applied nationally would lead to an additional 180,000 cases of hospital-acquired pneumonia and 33,000 preventable deaths annually (based on estimated mortality rate of 18%) [96].

For this reason assessing confounders and eliminating heterogeneity within studies is essential to ascertain the extent of this potential interaction.

Confounding and Heterogeneity

Confounding in the association between PPI use and CAP was the focus of a retrospective claims-based cohort study which used a "falsification approach" to determine if PPI use was implausibly associated with other common medical conditions [97].

This was performed by reviewing the prescription data for 6 US private insurance plans and comparing PPI use versus non-use to CAP, osteoarthritis, chest pain, urinary tract infection, deep vein thrombosis, skin infection, and rheumatoid arthritis.

Higher adjusted rates of CAP ($p<0.001$) and all other conditions ($p<0.001$) were seen for the quarters in which PPI users filled their prescription compared to those quarters where no prescription was filled [97]. This suggests that the observed association between PPI use and CAP may be do to confounding, as PPI use is also implausibly associated with common medical conditions.

Heterogeneity in the association between statins, ACE inhibitors, and PPIs, in the risk for pneumonia was also the focus of a recent study applying a common protocol to case control studies in different data settings; the authors concluded that case-control selection and methods of exposure ascertainment induce bias that can not be adjusted for and that may explain some of the heterogeneity seen in these studies [98].

Clinical Summary

Numerous studies and meta-analyses have shown an apparent association between PPI use and an increased risk of pneumonia. In particular, this risk is most pronounced in the short-term setting with no significant association noted for long-term use [89-93]; however, the results of these studies should be interpreted with caution, as a significant level of heterogeneity precludes any definitive relationship [92-94].

Further, there is the possibility for a protopathic bias to explain this temporal association. Lastly, recent studies have suggested the potential for significant confounding in this relationship, as PPI use has been implausibly associated with common medical conditions in addition to PPI use [97].

As such clinicians should be aware of the potential short-term association of PPI use with pneumonia, however, continue to use PPI where appropriately indicated.

Clostridium Difficile Infection

Of all the implicated risks of PPI use, the potential association between PPIs and the development of *Clostridium difficile* infection (CDI) is one of the most controversial. An early hypothesis for the mechanism of this increased risk was that alterations of gastric acid would limit the killing of *C. difficile* spores in gastric contents; however, experimental results have shown that most *C. difficile* spores are acid resistant, as gastric acid did not significantly alter the concentration of viable spores for 6 different *C. difficile* strains [99]. More recently, attention has been given to the role of acid suppression in the alteration of the gut microbiota and potentially decreasing the barriers for *C. difficile* colonization of the gut. This includes the possibility that PPIs may increase the ability of *C. difficile* spores to convert to the vegetative form within the lumen of the GI tract due to microbial alteration. Loss of colonic diversity is a common feature of CDI [100, 101] and has been evidenced to occur in an animal model of healthy dogs given PPIs [102]; more recently, evidence has shown that PPI administration for 28 days reduces microbial diversity in health human subjects [103]. Overall, the mechanisms behind which PPIs may increase the risk of CDI are poorly understood, with numerous potential interactions and limited data supporting them. Making matters more confusing, there are hundreds of available studies on *Clostridium difficile* infection and its potential risk factors, spanning a variety of healthcare settings and utilizing a variety of study designs. This section will seek to discuss the most current evidenced based research and offer a clinical summary integrating the available information.

In 2007, a systematic review (12 studies, 2,948 patients) was published concerning the use of acid suppressive medication and the risk of enteric infections. A major finding of this study was an increased risk for CDI among patients taking acid suppressive medication (OR 1.94, 95% CI, 1.37-2.75) [104]. While this study was subject to marked heterogeneity, it prompted continued and focused research on the potential association between acid suppressive medications, in particular PPIs, and the risk for CDI. Since that time, a number of meta-analyses have been conducted using a variety of statistical measures to clarify this potential association. While some of the most recent studies still have some limitations, their composite findings offer the best evidence for this association, as it is unlikely that any prospective trails can be completed given the ethical considerations and patient risk.

The first meta-analysis (23 studies, 300,000 patients) to look at the relationship between PPI use and CDI was published in 2012, identifying an overall risk estimate of 1.69 (95% CI 1.40-1.97) [105]. Importantly, there was a high level of heterogeneity in these studies (i^2 = 91.93%) and also visual evidence of asymmetry on funnel plot. When using an adjusted estimate and a trim-and-fill model to account for this, the adjusted risk was 1.26 (95% CI 1.03-2.24, p = 0.025) [105]. Shortly following this study, a second meta-analysis of 313,000 patients from 42 observational studies also identified an significant risk for PPIs and CDI (OR 1.74, 95% CI 1.47-2.85, p <0.001) [106]; however, similar to the first meta-analysis, this also was subject to considerable heterogeneity (i^2 = 85%), which could not be fully explained in subgroup analysis and remained when using only studies with adjusted risk estimates. Interestingly, subgroup analysis did show that H2RAs were associated with a lower risk for CDI than PPIs (OR 0.71, 95% CI 0.53-0.97) and also that PPIs in combination with antibiotics posed a significantly higher risk for CDI above PPIs alone (OR 1.96, 95% CI 1.03-3.70) [106]. Supporting a positive association was yet another meta-analysis published around the same time [107]. In this review of 202,965 patients from 30 studies, PPI use was associated with an OR 2.15 (95% CI 1.81-2.55, P <0.00001), however, again there was significant heterogeneity (i^2 = 89%) [107]. Overall, the presence of heterogeneity limits the generalization of this data and brings up concern for potential confounding and bias.

Given the considerable heterogeneity present in these meta-analyses, Tleyjeh et al. conducted a systematic review and meta-analysis using a novel regression based method to control for publication bias [108]. Overall, the pooled risk estimate revealed an OR of 1.65 (95% CI 1.47-1.85) albeit with considerable heterogeneity (i^2 = 89.9%). When exploring the sources of the heterogeneity, the authors observed that studies using interviews to ascertain PPI use had lower average risk estimates than those using medical records. Additionally, studies using adjusted risk estimates noted a higher risk than those using unadjusted estimates. In this meta-analysis, publication bias was noted both through visual asymmetry of the funnel plot and for Egger's test for publication bias (P = 0.001) [108]. When using the fitted regression-based model, the adjusted averaged effect estimate for PPI use in CDI was 1.51

(95% CI, 1.26-1.83) [108]. Using an incidence of CDI at 14 days after hospital admission of 42/1,000 for patients receiving antibiotics and 5.4/1,000 for patients without antibiotics, the adjusted effects model predicts a number needed to harm (NNH) of 50 and 367, respectively. Extrapolating this to the general population where the incidence of CDI at 1 year is 48/100,000, the NNH is 3925 [108].

In all there are several important discussion points from this systematic review and meta-analysis. First, there is considerable heterogeneity and publication bias among the available studies on the risks of PPIs for CDI. Even when adjusting for these aspects, these results constitute very low quality evidence using the GRADE system. Second, applying this increased risk to a population model reveals a NNH of 3925 for the general population. The large NNH supports the use of PPIs in the general population for whom an appropriate indication for PPI use is present. Even among hospitalized patients, the NNH for patients not receiving antibiotics is 367.

This finding should prompt clinicians to be aware of the potential association between CDI and PPI use, however, use clinical judgment as the clinical benefit of PPIs in many patients likely outweighs the risk of CDI.

Recently, there have also been several studies looking at particular subgroups of patients taking PPIs and their respective risk for CDI. Concerning the relationship for PPIs and increased severity of CDI, Khanna et al. performed a population-based study of 385 patients comparing demographic data and outcomes, including severe, severe-complicated, recurrent CDI, and treatment failure between patients receiving or not receiving long term acid suppression medication [109]. While univariate analysis showed that patients on acid suppressive medications were more likely to be significantly older (69 vs 56 years, $P = 0.006$) and more likely to have severe (34.2% vs 23.6%, $P = 0.03$) or severe-complicated (4.4% vs 2.6%, $P = 0.006$) CDI, when using multivariate analysis controlling for age and comorbidities, no association remained with for acid suppression and severity of CDI (severe CDI, $P = 0.23$; severe-complicated CDI $P = 0.07$) [109]. Additionally, there was no association with treatment failure or CDI recurrence in relation to acid suppression ($p = 0.35$ and $p=0.08$, respectively).

Among critically ill patients, a retrospective review of 3,286 patients identified PPI use as an independent risk factor in multivariate analysis for the 110 patients who developed CDI (OR 3.11, 95% CI 1.11-8.74, $p <0.05$) [110]. In this study, 55.6% of patients were receiving PPIs in contrast to 5.8% receiving H2RAs; importantly this suggests there may be a bias whereby PPIs are prescribed to “more ill” patients and H2RAs to those “less ill” patients.

An increased risk of recurrent CDI has also been suggested by some early studies [106]. This association, however, was not supported by the most recent and comprehensive study specifically addressing the role of PPIs in recurrent CDI. Freedberg et al. identified 894 inpatients with incident CDI, 23% of whom went on to develop recurrent CDI [111].

Among these patients, PPI use was not associated with CDI recurrence (HR 0.82, 95% CI 0.58-1.16); rather, black race (HR 1.66, 95% CI 1.05-2.63), increased age (HR1.02, 95 CI% 1.01-1.03), and increased comorbidities (HR 1.09, 85% CI 1.04-1.14) were associated with CDI recurrence. Further, the association remained insignificant when accounting for increased duration or dose of PPI (HR 0.87, 95% CI 0.60-1.28). While there was an association between PPI use and increased mortality in univariate analysis, this was no longer significant when using multivariate analysis.

Clinical Summary

Overall, there is conflicting evidence to support a definitive relationship between PPI use and the development of CDI. Despite this, a number of meta-analyses have demonstrated a significant association even when attempting to control for risk factors [105-108]. These studies should be interpreted with some caution, however, as there was considerable heterogeneity and evidence of publication bias, constituting a very low-grade quality of evidence.

Further, even with the positive association, the NNH for the general population is 3925 and for hospitalized patients not on antibiotics the NNH is 367 [108]. Lastly, neither the Infectious Disease Society of America (IDSA) nor the American College of Gastroenterology (ACG) recommend routine discontinuation of PPI use in their clinical guidelines for the diagnosis and treatment of CDI [112, 113].

As such, the benefits of PPI use in patients for whom there is an appropriate indication for acid suppression likely outweighs the risk of CDI, and PPI therapy should not be discontinued on the basis of a potential risk for CDI.

Small Intestinal Bacterial Overgrowth

The role of the gut microbiota in relation to gastrointestinal disorders is a hot topic that has recently been gaining explosively broad attention. Small intestinal bacterial overgrowth (SIBO) occurs when normal colonic bacterial colonize the small intestine. Evidence from animal models has suggested that

PPI use is associated with intestinal overgrowth, in particular anaerobic bacteria [114]. This is of particular importance as a higher incidence of spontaneous bacterial peritonitis (SBP) has been observed in cirrhotic patients with SIBO than those without; however, any causal relationship remains to be established [115].

Further complicating matters, PPIs have also been shown to alter normal gut motility, evidenced by delayed gastric emptying and increased antral motility at therapeutic doses [116]. While the clinical evidence concerning the role of PPIs and the risk of SBP will be addressed in the next section, it is important to highlight complex relationships that exist within the gut microbiota and the potential for confounding between them, in particular when considering the risk for SIBO.

To date, results from clinical studies are difficult to interpret, as there are technical challenges in establishing the diagnosis of SIBO. The gold standard for diagnosis of SIBO is a duodenal-jejunal aspirate and culture showing $>10^5$ colony forming units; however, in addition to being a poorly reproducible test, this presents a number of practical limitations when used in research. Instead, the glucose hydrogen breath test (GHBT) is a preferred alternative measure in most research settings, although it is subject to the use of varying thresholds and sensitivities. Additionally, there are conflicting results between the available studies that show variation not only by method of diagnosis, but also appear to vary across populations. This is not unexpected, as recent evidence has shown variability in an individual's gut microbiota, influenced by both genetic and environmental conditions [117]. In addition, certain medical conditions, especially Irritable Bowel Syndrome (IBS), predispose to the development of SIBO and have the potential to confound the results from some of these studies [118].

One of the initial reports highlighting an association between PPI use and SIBO can from a study of 450 consecutive patients undergoing GHBT. Two hundred patients with GERD using PPI therapy for a median duration of 36 months were compared to 200 patients with IBS in absence of PPI therapy for at least 3 years and also to 50 healthy controls [118]. Small intestinal bacterial overgrowth was found in 50% of patients on PPI therapy, compared to 24.5% of IBS patients, and 6% of healthy controls, indicating a statistically significant difference in those patient taking PPIs ($p = <0.001$) [118]. Further, the prevalence of SIBO increased after 1-year of PPI therapy and continued to worsen in symptom severity at 36 months. Importantly, treatment with rifaximin 400mg 3 times daily for 14 days successfully eradicated 87% of SIBO in the PPI group; there was not a statistically significant difference in eradiation rates based on duration of PPI therapy (93% for PPI less than 12 months and 86% in those longer than 12 months) [118]. This is an important point as it suggests SIBO resulting from PPI use is reversible with treatment regardless of the duration of PPI therapy.

Results of this study were supported by a smaller study of 42 patients with nonerosive reflux disease without bowel symptoms on PPI therapy (esomeprazole 20mg BID) for 6 months [119]. After 8 weeks of PPI therapy there was a significant increase in bloating and flatulence ($p < 0.001$ and $p < 0.01$, respectively); this continued to increased over the trial and at 6 months there was a significant increase in bloating, flatulence, abdominal pain, and diarrhea compared to baseline ($p < 0.0001$, $p < 0.0001$, $p < 0.001$, and $p < 0.01$, respectively) [119]. Importantly, 11 of 42 patients had a positive GHBT at 6-months compared to none at baseline ($p < 0.05$). Similarly, another recent study found a significantly positive association for PPI use and SIBO in a retrospective review of 150 patients receiving duodenal aspirates and cultures (OR of 2.46, $p = 0.008$); among the PPI users 62% (40/65) developed SIBO as diagnosed by a cutoff of $>10^5$ colony forming units [120]. Combined, these studies suggest that prolonged PPI treatment may produce bowel symptoms and is positively associated with an increased risk for SIBO.

In contrast, two subsequent studies have found no significant association of PPI use with SIBO. In a retrospective case review of 675 patients who received duodenal aspirates, 10% of patients using PPIs had an abnormal result compared to 6% not taking a PPI, a non-significant finding; however, PPI use was associated with bacterial growth not meeting criteria for SIBO in both IBS patients (OR = 4.7, 95% CI 1.2-18.7, $p = 0.03$) and non-IBS patients (OR 2.7, 95% CI 1.4-4.9, $p < 0.01$) [121]. Similarly, a retrospective chart review of 1191 patients undergoing GHBT looked for the presence of risk factors for SIBO. In total, 566 (48%) of patients were taking a PPI, however, PPI use was not significantly associated with SIBO using any of the diagnostic criteria, even after controlling for other potential confounders [122]. Rather, GHBT positivity was only associated with older age (OR 1.03, 95% CI 1.01-1.04) and antidiarrheal use (1.99, 95% CI 1.15-3.44) [122].

Most recently, a systematic review and meta-analysis (11 studies, 3134 patients) found that PPI use was associated with SIBO risk, but only when the diagnosis was made by duodenal or jejunal aspirate (OR = 2.28, 95% CI 1.24-4.21); however, in addition to notable heterogeneity (i^2 83.7, $p < 0.001$), funnel plot revealed 4 outlying studies, which may have introduced significant publication bias [123]. When accounting for this potential publication bias, the adjusted OR showed no relationship between the risk of SIBO and PPI use

(OR 1.44, 95% CI 0.78-2.67). When divided based on diagnostic method, a significant relationship was found for duodenal or jejunal aspirate and culture (OR 7.59, 95% CI 1.81-31.89) but not for GHBT (OR 1.93, 95% CI 0.69-5.42).

Another important subgroup analysis highlighted a significant association in studies using self-controls or pretest-post test design (OR 9.607, 95% CI 1.65-56.01) compared to no significant association in those using PPI nonusers as a control (OR 1.66, 95%CI 0.87-3.16). This observation suggests the potential role for an individual's gut microbiota to act as an influential factor in the development of SIBO. It also brings up the possibility of unrecognized confounding within these studies, in particular patient age and unadjusted comorbidities.

Clinical Summary

While there does appear to be a pathophysiological basis for an increased risk of SIBO in patients taking PPI, the current clinical evidence is controversial and prohibits any definitive conclusion. This is particularly true given the lack of randomized control trials, variability in diagnostic measures, and potentially unrecognized confounding. Still, clinicians should be aware of the potential risk for bowel symptoms and SIBO in patients receiving long-term PPI therapy. In patients who do develop SIBO, rifaximin 400mg three times daily for 14 days has been shown to effectively eradicate the majority of SIBO regardless of the duration of PPI therapy [118].

Spontaneous Bacterial Peritonitis

A number of studies have suggested that PPI use is associated with an increased risk of spontaneous bacterial peritonitis (SBP) in patients with cirrhosis [124-128]. The etiology of this increased risk for SBP is centered on a decreased ability of host defenses to inhibit the proliferation of bacteria, since gastric acid plays an essential role in decontamination of the stomach and small intestine. Evidence suggests that PPIs can also alter neutrophil function by both inhibiting the expression of endothelial cell adhesion molecules [129] and impairing the production of reactive oxygen intermediates [130]. Further, PPIs may increase the risk for small intestinal bacterial overgrowth (SIBO), which is a risk factor for SBP as it increases the chance for bacterial translocation into the lymph nodes and subsequent spread to the ascites [131, 132]. Despite this pathophysiological basis for an increased risk of SBP in cirrhotic patients on PPIs, the clinical evidence is not so clear.

Initial studies concerning the role of PPIs in SBP revealed conflicting results.

Campbell et al. performed a retrospective case control study of 116 consecutive cirrhotic patients with ascites undergoing diagnostic paracentesis to determine the role of PPI use and the risk for SBP [133]. A total of 32 patients were diagnosed with SBP. PPI use was similar between the two groups, 41% in the SBP group versus 36% without. PPI use was not significantly associated with the risk for SBP in multivariable analysis (OR 1.22, 95% CI 0.52-2.87) [133]. Further, the risk remained insignificant when the definition of acid suppression was expanded to include the use of H2RAs. Importantly, patients diagnosed with SBP were found to have a higher MELD score (Model for End-Stage Liver Disease) (23 versus 18, $p = 0.002$), increased INR (international normalized ratio) (1.8 versus 1.5, $p = <0.001$) and increased bilirubin (4.1 versus 2.6, $p=0.01$), indicating that they had more advanced liver disease [133].

Attempting to improve on this study design, Bajaj et al. conducted a retrospective case-controlled study of 70 cirrhotic patients with paracentesis-proven SBP [124]. Given that the study by Campbell et al. showed patients with SBP had more severe liver impairment [133], Bajaj et al. matched each SBP patient with a comparable cirrhotic (both age and Child's class) with ascites admitted for conditions other than SBP [124]. Additionally, they excluded patients with recent upper GI bleeding or prior antibiotic use, which may have represented confounders in the earlier study. Among the patients diagnosed with SBP, there was a significantly greater use of PPIs for at least 1 week prior to admission when compared to patients without SBP (69% vs 31%). Multivariate analysis showed that PPI use was independently associated with SBP (OR 4.31, 95% CI 1.34-11.7) [124].

As a result, the authors concluded that PPI therapy is associated with SBP in patients with advanced cirrhosis.

This association also appears to hold true across different populations. In South Korea, Choi et al. found a significant association between PPI use and the development of SBP [125]. In this retrospective review of 176 patients admitted for diagnostic paracentesis, 83 patients were found to have SBP. Using multivariate analysis, independent risk factors for SBP were Child-Pugh Class C (OR 2.89, 95% CI 1.44-5.79), MELD score ≥ 20 (OR 3.54, 95% CI 1.56-10.85), and PPI use (OR 3.44, 95% CI 1.16-10.19) [125]. Among patients within the PPI group, median duration of PPI use was 91 days in the SBP group versus 29 days in the non-SBP group (p = 0.094); the authors suggested this borderline significance highlights a potential association between length of PPI exposure and risk for SBP. Despite the significant finding for PPI use as a risk for SBP in this particular study, the percentage of PPI use was much lower than other studies (18.1% in the SBP group versus 6.5% in the non-SBP group). This ranges 20-50% less than other studies [124, 126, 127, 133] and when combined with the overall small number of subjects in these subgroups (15 PPI users in the SBP group and 6 in the non-SBP group) limits the application of the results to the general population.

More recently in Japan, Miura et al. found an independent risk for PPI use and the development of SBP in 65 cirrhotics with ascites (OR 6.41, 95% CI 1.16-35.7, p=0.033) [134]. These authors do note several limitations to their study including a small sample size and potential for confounding conditions. An important finding in this study, however, was again that both the MELD score and Child-Pugh classifications were higher in patients with SBP, indicating more advanced liver disease.

Additionally when monitoring patient survival, only MELD score, not PPI use or the presence of SBP, was independently associated with overall mortality. A major limitation for all the above-mentioned studies is a lack of subgroup analysis within the PPI groups comparing short-term to long-term PPI use. The study by Goel et al., however, closely looked at these subgroups among 130 hospitalized cirrhotic patients and found two particularly interesting results [127]. First, patients who had not taken a PPI within the past 90 days had a 70% less likelihood of having SBP than those who had taken a PPI in the last 7 days; second, patients who had PPI use in the previous 90 days, but not within 7 days had a 79% less likelihood of having SBP compared to those who had PPI use within the past 7 days [127]. This study also found that the risk of SBP in patients treated with H2RAs in the past 30 days was nearly identical to the risk for PPIs [127].

Overall, the results of this study suggest that it may be recent (<7 days) use of acid suppressive medication, either PPI or H2RA, that is primarily responsible for the observed risk of SBP in these numerous studies.

Stronger evidence comes from a systematic review and meta-analysis (4 studies, 772 patients) that found a positive association between the use of PPI and development of SBP (OR 2.77, 95% CI 1.82-4.23) [128]. The major strengths of this study were the pooled analysis results, which allowed for a much larger sample size and tighter confidence intervals. Overall, these results encourage clinician awareness of the potential association between PPI use and the development of SBP.

Despite this, two limitations to the study should be noted. First, only 2 of the 4 studies included in this paper were peer-reviewed manuscripts. Second, over half of the patients (403/772) came from a single study for which only an abstract is available [135].

These limitations do not weaken the strength of the evidence from this study; rather, they should highlight the challenges of establishing definitive relationships in a disease as complicated as cirrhosis, which has numerous potential cofounders. A particular concern for which would be alterations of metabolism secondary to liver dysfunction. This can increase the half-lives of PPIs by 4-8 hours, potentially increasing the risk of accumulation and toxicity [136]. Further, there are a number of practical limitations prohibiting quality prospective trials on this topic, making our reliance on these retrospective studies critical.

Since the publication of this meta-analysis, two large retrospective reviews have additionally suggested that PPI use increases the risk of SBP in cirrhotics. The first study was a multicenter cohort analysis including 1140 cirrhotic patients, 533 identified as having SBP; among these patients, the use of PPIs and HR2A were both associated with the development of SBP [137]. Adjusted odds ratios for each subgroup were: PPI use with in 7 days (aOR 2.10, 95% CI 1.33-3.33, $p = 0.001$), PPI use within 8-30 days of admission (aOR 2.87, 95% CI 1.30-6.31, $p = 0.009$), H2RA use within 7 days (aOR 6.00, 95% CI 2.70-13.36, $p < 0.001$), H2RA within 8-30 days of admission (aOR 3.84, 95% CI 1.02-14.37, $p = 0.046$). The authors also looked at mortality after SBP using a multivariate analysis, finding PPI use within 30 days (aOR 1.96, 95% CI 1.19-3.23, $p = 0.008$), higher admission MELD score (aOR 1.05, 95% CI 1.03-1.08, $p < 0.001$), and hepatocellular carcinoma (aOR 1.85, 95% CI 1.26-2.73, $p = 0.002$) were all significant risk factors [137]. The second study performed at a single medical center included 1965 cirrhotic patients, finding that among the 32.9% of patients who were PPI users, there was an increased incidence of SBP compared to non-PPI users (10.6% vs 5.8%, respectively, $p = 0.02$) [138]. Even when using propensity score matching to account for disease severity, this relationship held true with an incidence of 10.8% in PPI users compared to 6.0% for non-PPI users ($p = 0.038$). An additional finding within the PPI group was that after controlling for gender, Child-Pugh score, and serum sodium, the dose of PPI was an independent risk factor for SBP. Lastly there was limited value in the duration of PPI use as a risk factor, as the mean duration of PPI use in patients with SBP was 74.4 ± 88.8 days [138].

Finally, exacting the role of PPIs in the development of SBP is further complicated by the possibility that PPIs, themselves, are a risk factor not only for the severity of cirrhosis and also increased mortality.

This was suggested from a recent prospective trial enrolling 272 cirrhotic patients, which found that PPI use, in addition to MELD score, hepatocellular carcinoma, and hepatic decompensation, was an independent predictor of mortality in multivariate analysis (HR 2.33, 95% CI 1.26-4.30, $p = 0.007$) [139]. Overall PPI use was 78.3% (213/272), with greater than 50% of patients taking a PPI for symptomatic relief. While the results of this study clearly show an association between severity of cirrhosis and PPI use, it can not be established if PPIs are the cause or simply a common therapeutic modality seen in patients with more advanced cirrhosis. Further investigation in this topic will not only help to identify the risk of PPI use in cirrhosis, but also help to explain some of the potential confounding that is apparent when discussing the potential increased risk for SBP among PPI users.

Clinical Summary

While initial studies suggested a minimal risk of PPI use associated with the development of SBP in cirrhotic patients with ascites, several recent studies have all supported an associated risk [126-128, 137, 138]. There are still certain limitations within these studies, namely many with small sample sizes and retrospective designs.

Additionally, there are a number of confounders in these cirrhotic patients that may not have been accounted for, in particular, when the majority of studies indicate the SBP patients tend to have more advanced liver disease. Despite this, it remains clear that clinicians should scrutinize the use of PPIs in this patient population and ensure that patients taking these medications have an appropriate indication for their use.

Travelers' Diarrhea

A long-standing and controversial association of PPI therapy is an increased risk for enteric infections, in particular, traveller's diarrhea. This specific risk, however, has not been formally targeted in any studies of travelers [140].

One study did review the risk for travelers' diarrhea among a cohort of 743 travellers to endemic areas, finding the use of acid suppressive medication was not significantly associated with an increased risk for travelers' diarrhea in subgroup analysis (OR 6.9, range 0.7-67.4) [141]. Despite this, two systematic reviews have found an increased risk of enteric infections in patients taking PPIs, especially for *Salmonella* and *Campylobacter* [104, 142]. This first review (6 studies, 11,280 patients) noted an overall increased risk for enteric infections among patients taking acid suppression (OR 2.55, 95% CI 1.53-4.26), albeit with significant heterogeneity ($p<0.0001$) [104].

In subgroup analysis, this association was greater for those taking PPIs (OR 3.33, 95% CI 1.84-6.02) than H2RAs (OR 2.03, 95% CI 1.05-3.92) [104]. More recently, a second systematic review found that PPIs increased susceptibility to *Salmonella, Campylobacter jejuni,* invasive strains of *Escherichia coli*, *Vibrio cholerae,* and *Listeria* [142]. Of these, *Salmonella* and *Campylobacter* showed particularly increased risk with adjusted relative risk ranges of 4.2-8.3 (two studies) and 3.5-11.7 (four studies), respectively. The authors propose this increased may be related to a mechanism whereby bacterial species have increased survival at a higher pH; experimental data shows that the *Salmonella* strains *S. paratyphi*, *S. enteritis, and S. enterica servoa Trphyimurium* all have limited survival at pH <3.0-3.5 with increased survival at pH >3.5-4.0 [143, 144]. Additionally, the authors postulate that PPI effects on neutrophil functions and intestinal permeability increase the risk of infection [142]. In contrast, *E. coli* appears to be more acid stable in medium and also have pH dependent acid tolerance strategies for survival in gastric contents, yet requires a minimum pH of 4.4 to support bacterial proliferation [145, 146]. Further, evidence from healthy adults volunteers showed diarrhea from *E. coli* only occurred after neutralization of gastric acid [147]. Taken together, these points suggest an increased risk for infection among those with a decreased gastric acidity.

Clinical Summary

There is currently limited data to support a definitive risk for travelers' diarrhea among PPI users. While there is evidence to support an increased risk of enteric infections related to acid suppressive medications [104, 142], this has not been studied specifically in travelers.

The International Society of Travel Medicine does, however, consider the use of acid suppressive medication to be a risk factor for enteric infections and recommends that patients taking these medications be considered for chemoprophylaxis when traveling to high-risk endemic areas [148].

V. Other Reported Adverse Effects

Interstitial Nephritis

Acute interstitial nephritis (AIN) is a pattern of renal injury marked by inflammation and edema of the renal interstitium, sometimes involving the renal tubules, due to humoral and cell-mediated hypersensitivity.

AIN is seen in approximately 1% of renal biopsies evaluated for hematuria or proteinuria and has been estimated to be responsible for 5-15% of acute renal failure cases [149]. AIN is important to recognize early as there are a number of severe sequelae including hospitalization, renal biopsy, high dose corticosteroids and/or immunosuppressants, and the development of chronic kidney disease [150]. Most commonly, the etiology of AIN is thought to be drug-induced, infection-associated, or immune mediated.

Among cases of drug-induced AIN, nonsteroidal anti-inflammatory drugs (NSAIDs), antibiotics, and diuretics are the most cited culprits. There are, however, a multitude of other medications associated with AIN, including the H2RAs, famotidine and ranitidine [149]. More recently, PPIs have been associated with AIN following several case reports and a systematic review.

A recently published case series from India described four cases of PPI-induced AIN (esomeprazole, omeprazole, two with pantoprazole) [151]. Symptoms of vomiting, loin pain, and oliguria were noted an average of 4 weeks after drug therapy (range 10 days to 8 weeks).

Upon recognition for AIN, the PPI was stopped in each case and steroid therapy initiated with two patients requiring hemodialysis; complete recovery was noted in two patients at 1 year of follow up, while the other two patients showed partial recovery at 4-6 months of follow-up.

The most comprehensive data to date comes from a systematic review of 60 case reports of PPI-associated AIN (59 biopsy proven cases). Overall, the clinical presentation of these patients was non-specific, most commonly nausea (30%), emesis (30%), malaise (23%), or asymptomatic (10%). Mean time to development of AIN was 13 weeks with a range of 2 weeks to one year. The PPIs used were omeprazole (47 cases), pantoprazole (6 cases), esomeprazole (3 cases), lansoprazole (2 cases), and rabeprazole (2 cases). In all cases, PPI therapy was stopped upon recognition of AIN with 20% of cases requiring corticosteroid treatment and three cases requiring dialysis; average time to recovery was 35.5 weeks.

Given the widespread use of PPIs worldwide, the authors concluded that PPI-related AIN is a rare, idiosyncratic occurrence without enough current evidence to establish any causal effect [152]. This is supported by a retrospective case-control study of 68 AIN cases, which found PPI use was associated with an OR of 1.05 (95% CI 0.97-1.14) for development of AIN; as such the authors conclude that PPI exposure may increase the odds of AIN, but this result is not definitive and should be confirmed with larger studies to increase the statistical power [150].

Clinical Summary

Despite its rare and idiosyncratic occurrence, PPI-induced acute interstitial nephritis is a real clinical entity with the potential for severe sequelae. As such clinicians should have a high level of suspicion to detect AIN early in its presentation, particularly if patients were recently started on a PPI.

Methotrexate

In December of 2011 the FDA issued a safety warning concerning the concomitant administration of high-dose methotrexate with PPIs and the potential for methotrexate toxicity. This was based primarily on two case reports of delayed methotrexate elimination and also pharmacokinetic studies suggesting a prolonged elevation in the levels of methotrexate and/or its metabolite, 7-hydroxymethotrexate, when administered with PPIs [153]. This delayed elimination of methotrexate is postulated to occur as a result of competition for certain transporter proteins, namely the breast cancer resistance protein (BCRP), between methotrexate and PPIs [154]; however, the clinical significance of this interaction has yet to be determined.

An initial finding supporting an interaction between PPIs and methotrexate came from a study of 76 patients receiving high-dose methotrexate, which showed a 27% and 39% decrease in clearance for both methotrexate and 7-hydroxymethotrexate, respectively, among patients receiving concomitant PPI therapy; additionally, patients on concomitant PPIs had significant elevations in 24 and 48 hour plasma concentrations of both methotrexate and 7-hydroxymethotrexate [155]. Supporting this finding were results from data of 171 cycles of methotrexate among 74 patients, which showed co-administration of PPIs as a risk factor for delayed methotrexate elimination (OR 2.65, 95% CI 1.03-6.82) [156]. Importantly, PPI administration was shown to inhibit the BCRP-mediated transport of methotrexate, however, in vitro analysis showed this occurred at PPI levels 50-200 times the therapeutic range and was also not seen in every patients within the study. As such, the drug interaction cannot fully be explained by PPI inhibition of BRCP. The authors suggest that there may be genetic factors influencing this interaction, namely genetic polymorphisms in CYP2C19 or BRCP, which may alter PPI levels or methotrexate transport [156].

Interestingly, a recent retrospective review of 56 patients undergoing a total of 201 cycles of methotrexate found that while PPIs increased median methotrexate levels ($p = 0.013$), there was no significant difference in the proportion of patients experiencing delayed elimination at 24 or 72 hours ($p=0.765$) [157]. Additionally, when controlling for multiple cycles of methotrexate per patients, concomitant use of PPI was not a significant predictor for methotrexate levels ($p=0.969$) [157]. As a result the authors of this study conclude that any interaction between PPIs and methotrexate, if present, is likely to be small. Finally, it is important to note that a review of the FDA Adverse Event Reporting System yielded a number of case reports in which an H2RA was substituted for a PPI without any methotrexate toxicity reported [158].

Clinical Summary

There is evidence to suggest a potential mechanism whereby PPI co-administration with high-dose methotrexate may lead to delayed drug elimination and methotrexate toxicity.

Clinical evidence from retrospective reviews, however, offers conflicting evidence on the prevalence and significance of this association.

As such clinicians should be aware of this potential interaction and monitor patients closely for any signs of toxicity. Further, given that there are no reported interactions with H2RAs, physicians should consider switching therapy prior to high-dose methotrexate therapy.

Conclusion

To date there have been a number of reported complications from PPI use, ranging from rare idiosyncratic effects to more prevalent complications, including alteration of pharmacokinetics and increased rates of infection. Table 2 reviews the clinical summary for each of the discussed risks. Despite these reported risks, the current medical literature offers conflicting evidence on nearly all topics. This is largely related to the retrospective design of most investigations and heterogeneity apparent in many meta-analyses. Given the overprescription of PPIs in the general population, clinicians should scrutinize the use of PPIs, keeping them restricted to patients with appropriate indications. In many of the patients for whom there is an appropriate indication, the benefits of PPI use likely outweigh the potential for adverse events. Continued research will undoubtedly serve to distinguish the role of PPIs from potential confounders and also clarify the true risk for many of these interactions.

Until then, clinicians should continue to discuss the risk-benefit profile of PPIs with patients and stop their use in patients for whom they are not needed.

Conflicts of Interest: Edward C Oldfield IV: no disclosures
David A Johnson

References

[1] Rabin, R. Combating Acid Reflux May Bring Host of Ills. *The New York Times*. 25 June 2015 [http://well.blogs.nytimes.com/2012/06/25/combating-acid-reflux-may-bring-host-of-ills/].

[2] Force, R. W., Meeker, A. D., Cady, P. S., Culbertson, V. L., Force, W. S., Kelley, C. M. Increased Vitamin B12 Requirement Associated with Chronic Acid Suppression Therapy. *Ann. Pharmacother.* 2003; 37: 490–493.

[3] Andrès, E., Loukili, N. H., Noel, E., Kaltenbach, G., Ben Abdelgheni, M., Perrin, A. E., Noblet-Dick, M., Maloisel, F., Schlienger, J. L., Blicklé, J. F. Vitamin B12 (cobalamin) deficiency in elderly patients. *CMAJ* 2004: 251–259.

[4] Dharmarajan, T. S., Kanagala, M. R., Murakonda, P., Lebelt, A. S., Norkus, E. P. Do Acid-Lowering Agents Affect Vitamin B12 Status in Older Adults?. *J. Am. Med. Dir. Assoc.* 2008; 9: 162–167.

[5] Rozgony, N. R., Fang, C., Kuczmarski, M. F., Bob, H. Vitamin B(12) deficiency is linked with long-term use of proton pump inhibitors in institutionalized older adults: could a cyanocobalamin nasal spray be beneficial?. *J. Nutr. Elder.* 2010; 29: 87–99.

[6] Den Elzen, W. P. J., Groeneveld, Y., de Ruijter, W., Souverijn, J. H. M., le Cessie, S., Assendelft, W. J. J., Gussekloo, J. Long-term use of proton

pump inhibitors and vitamin B12 status in elderly individuals. *Aliment. Pharmacol. Ther.* 2008; 27: 491–497.

[7] Lam, J. R., Schneider, J. L., Zhao, W., Corley, D. A. Proton pump inhibitor and histamine 2 receptor antagonist use and vitamin B12 deficiency. *Jama* 2013; 310: 2435–42.

[8] Zhang, H., DiBaise, J. K., Zuccolo, A., Kudrna, D., Braidotti, M., Yu, Y., Parameswaran, P., Crowell, M. D., Wing, R., Rittmann, B. E., Krajmalnik-Brown, R. Human gut microbiota in obesity and after gastric bypass. *Proc. Natl. Acad. Sci. US* 2009; 106: 2365–2370.

[9] Bezwoda, W., Charlton, R., Bothwell, T., Torrance, J., Mayet, F. The importance of gastric hydrochloric acid in the absorption of nonheme food iron. *J. Lab. Clin. Med.* 1978; 92: 108–116.

[10] Conrad, M., Schade, S. Ascorbic acid chelates in iron absorption: a role for hydrochloric acid and bile. *Gastroenterology*. 1968; 55: 35–45.

[11] Mowat, C., Carswell, A., Wirz, A., McColl, K. E. Omeprazole and dietary nitrate independently affect levels of vitamin C and nitrite in gastric juice. *Gastroenterology* 1999; 116: 813–822.

[12] Koop, H. Review article: metabolic consequences of long-term inhibition of acid secretion by omeprazole. *Aliment. Pharmacol. Ther.* 1992; 6: 399–406.

[13] Johnson, D. A., Peura, D. A., Vaezi, M. F. Concomittant Use of PPIs and Antiplatelet Therapy. *Gastroenterol. Hepatol.* 2011: 7–9.

[14] Tempel, M., Chawla, A., Messina, C., Çeliker, M. Y. Effects of Omeprazole on Iron Absorption: Preliminary Study Omeprazol ' un Demir Absorpsiyonundaki Etkileri: Ön Çalışma. 2013.

[15] Sarzynski, E., Puttarajappa, C., Xie, Y., Grover, M., Laird-Fick, H. Association between proton pump inhibitor use and anemia: A retrospective cohort study. *Dig. Dis. Sci.* 2011; 56: 2349–2353.

[16] Shikata, T., Sasaki, N., Ueda, M., Kimura, T., Itohara, K., Sugahara, M., Fukui, M., Manabe, E., Masuyama, T., Tsujino, T. Use of Proton Pump Inhibitors Is Associated With Anemia in Cardiovascular Outpatients. *Circ. J.* 2014; 79: 193–200.

[17] FDA Drug Safety Communication: Low magnesium levels can be associationed with long-term use of Proton Pump Inhibitor drugs [http://www.fda.gov/Drugs/DrugSafety/ucm245011.htm].

[18] Epstein, M., McGrath, S., Law, F. Proton-Pump Inhibitors and Hypomagnesemic Hypoparathyroidism. *N. Eng. J. Med.* 2006; 355: 1834–1836.

[19] Broeren, M. A., Geerdink, E. A., Vader, H., van den Wall Bake, A. Hypomagnesemia induced by several proton-pump inhibitors. *Ann. Intern. Med.* 2009; 151: 755–756.

[20] Mackay, J. D., Bladon, P. T. Hypomagnesaemia due to proton-pump inhibitor therapy: a clinical case series. *QJM* 2010; 103: 387–95.

[21] Schlingmann, K. P., Weber, S., Peters, M., Niemann Nejsum, L., Vitzthum, H., Klingel, K., Kratz, M., Haddad, E., Ristoff, E., Dinour, D., Syrrou, M., Nielsen, S., Sassen, M., Waldegger, S., Seyberth, H. W., Konrad, M. Hypomagnesemia with secondary hypocalcemia is caused by mutations in TRPM6, a new member of the TRPM gene family. *Nat. Genet.* 2002; 31: 166–170.

[22] Hou, J., Renigunta, A., Konrad, M., Gomes, A. S., Schneeberger, E. E., Paul, D. L., Waldegger, S., Goodenough, D. A. Claudin-16 and claudin-19 interact and form a cation-selective tight junction complex. *J. Clin. Invest.* 2008; 118: 619–628.

[23] Bai, J. P. F., Hausman, E., Lionberger, R., Zhang, X. Modeling and simulation of the effect of proton pump inhibitors on magnesium homeostasis. 1. Oral absorption of magnesium. *Mol. Pharm.* 2012; 9: 3495–3505.

[24] Hess, M. W., Hoenderop, J. G. J., Bindels, R. J. M., Drenth, J. P. H. Systematic review: hypomagnesaemia induced by proton pump inhibition. *Aliment. Pharmacol. Ther.* 2012; 36: 405–13.

[25] Biyik, M., Solak, Y., Ucar, R., Cifci, S., Tekis, D., Polat, I., Goktepe, M., Sakiz, D., Ataseven, H., Demir, A. Hypomagnesemia Among Outpatient Long Term Prot*on Pump Inhibitor. Am. J. Ther.* 2014; Epub. ahead. DOI: 10.1097/MJT.0000000000000154

[26] Van Ende, C., Van Laecke, S., Marechal, C., Verbeke, F., Kanaan, N., Goffin, E., Vanholder, R., Jadoul, M. Proton pump inhib*itors do not* influence serum magnesium levels in renal transplant recipients. *J. Nephrol.* 2014; Epub. ahead. DOI 10.1007/s40620-014-0105-9

[27] Lindner, G., Funk, G.-C., Leichtle, A. B., Fiedler, G. M., Schwarz, C., Eleftheriadis, T., Pasch, A., Mohaupt, M. G., Exadaktylos, A. K., Arampatzis, S. Impact of proton pump inhibitor use on magnesium homoeostasis: a cross-sectional study in a tertiary emergency department. *Int. J. Clin. Pract.* 2014; 68: 1352–7.

[28] Zipursky, J., Macdonald, E. M., Hollands, S., Gomes, T., Mamdani, M. M., Paterson, J. M., Lathia, N., Juurlink, D. N. Proton Pump Inhibitors and Hospitalization with Hypomagnesemia: A Population-Based Case-Control Study. *PLoS Med.* 2014; 11: e1001736.

[29] Danziger, J., William, J. H., Scott, D. J., Lee, J., Lehman, L., Mark, R. G., Howell, M. D., Celi, L. A., Mukamal, K. J. Proton-pump inhibitor use is associated with low serum magnesium concentrations. *Kidney Int.* 2013; 83: 692–9.

[30] Regolisti, G., Cabassi, A., Parenti, E., Maggiore, U., Fiaccadori, E. Severe Hypomagnesemia During Long-term Treatment With a Proton Pump Inhibitor. *Am. J. Kidney Dis.* 2010; 56: 168–174.

[31] Park, C. H., Kim, E. H., Roh, Y. H., Kim, H. Y., Lee, S. K. The Association between the Use of Proton Pump Inhibitors and the Risk of Hypomagnesemia: A Systematic Review and Meta-Analysis. *PLoS One* 2014; 9: e112558.

[32] FDA Drug Safety Communication: Possible increased risk of fractures of the hip, wrist, and spine with the use of proton pump inhibitors [http://www.fda.gov/Drugs/DrugSafety/PostmarketDrugSafetyInformationforPatientsandProviders/ucm213206.htm].

[33] Bo-Linn, G. W., Davis, G. R., Buddrus, D. J. An evaluation of the importance of gastric acid secretion in the absorption of dietary calcium. *J. Clin. Invest.* 1984; 73: 640–647.

[34] Nordin, B. E. C. Calcium and osteoporosis. *Nutrition* 1997; 13: 664–686.

[35] Eom, C., Park, S. M., Myung, S., Yun, J. M., Ahn, J. Use of acid-suppressive drugs and risk of fracture: a meta-analysis of observational studies. *Ann. Fam. Med.* 2011; 9: 257–67.

[36] Tuukkanen, J., Väänänen, H. K. Omeprazole, a specific inhibitor of H+-K+-ATPase, inhibits bone resorption in vitro. *Calcif. Tissue Int.* 1986; 38:123–125.

[37] Graziani, G., Como, G., Badalamenti, S., Finazzi, S., Malesci, A., Gallieni, M., Brancaccio, D., Ponticelli, C. Effect of gastric acid secretion on intestinal phosphate and calcium absorption in normal subjects. *Nephrol. Dial. Transpl.* 1995; 10: 1376–1380.

[38] Graziani, G., Badalamenti, S., Como, G., Gallieni, M., Finazzi, S., Angelini, C., Brancaccio, D., Ponticelli, C. Calcium and phosphate plasma levels in dialysis patients after dietary Ca-P overload: Role of gastric acid secretion. *Nephron* 2002; 91: 474–479.

[39] O'Connell, M. B., Madden, D. M., Murray, A. M., Heaney, R. P., Kerzner, L. J. Effects of proton pump inhibitors on calcium carbonate absorption in women: a randomized crossover trial. *Am. J. Med.* 2005; 118: 778–781.

[40] Hardy, P., Sechet, A., Hottelart, C., Oprisiu, R., Abighanem, O., Said, S., Rasombololona, M., Brazier, M., Moriniere, P., Achard, J. M., Pruna, A., Fournier, A. Inhibition of gastric secretion by omeprazole and efficiency of calcium carbonate on the control of hyperphosphatemia in patients on chronic hemodialysis. *Artif. Organs* 1998; 22: 569–573.

[41] Wright, M. J., Sullivan, R. R., Gaffney-Stomberg, E., Caseria, D. M., O'Brien, K. O., Proctor, D. D., Simpson, C. A., Kerstetter, J. E., Insogna, K. L. Inhibiting gastric acid production does not affect intestinal calcium absorption in young, healthy individuals: a randomized, crossover, controlled clinical trial. *J. Bone Miner. Res.* 2010; 25: 2205–2211.

[42] Hansen, K. E., Jones, A. N., Lindstrom, M. J., Davis, L. A., Ziegler, T. E., Penniston, K. L., Alvig, A. L., Shafer, M. M. Do proton pump inhibitors decrease calcium absorption?. *J. Bone Miner. Res.* 2010; 25: 2786–95.

[43] Schinke, T., Schilling, A. F., Baranowsky, A., Seitz, S., Marshall, R. P., Linn, T., Blaeker, M., Huebner, A. K., Schulz, A., Simon, R., Gebauer, M., Priemel, M., Kornak, U., Perkovic, S., Barvencik, F., Beil, F. T., Del Fattore, A., Frattini, A., Streichert, T., Pueschel, K., Villa, A., Debatin, K.-M., Rueger, J. M., Teti, A., Zustin, J., Sauter, G., Amling, M. Impaired gastric acidification negatively affects calcium homeostasis and bone mass. *Nat. Med.* 2009; 15: 674–81.

[44] Jefferies, K. C., Cipriano, D. J., Forgac, M. Function, structure and regulation of the vacuolar (H+)-ATPases. *Arch. Biochem. Biophys.* 2008: 33–42.

[45] Mizunashi, K., Furukawa, Y., Katano, K., Abe, K. Effect of omeprazole, an inhibitor of H+, K+-ATPase, on bone resorption in humans. *Calcif. Tissue Int.* 1993; 53: 21–25.

[46] Yang, Y.-X., Lewis, J. D., Epstein, S., Metz, D. C. Long-term proton pump inhibitor therapy and risk of hip fracture. *JAMA* 2006; 296: 2947–2953.

[47] Khalili, H., Huang, E. S., Jacobson, B. C., Camargo, C. A., Feskanich, D., Chan, A. T. Use of proton pump inhibitors and risk of hip fracture in relation to dietary and lifestyle factors: a prospective cohort study. *BMJ* 2012: e372–e372.

[48] Ngamruengphong, S., Leontiadis, G. I., Radhi, S., Dentino, A., Nugent, K. Proton pump inhibitors and risk of fracture: a systematic review and meta-analysis of observational studies. *Am. J. Gastroenterol.* 2011; 106: 1209–1218; quiz 1219.

[49] Targownik, L. E., Leslie, W. D., Davison, K. S., Goltzman, D., Jamal, S. A., Kreiger, N., Josse, R. G., Kaiser, S. M., Kovacs, C. S., Prior, J. C., Zhou, W. The relationship between proton pump inhibitor use and longitudinal change in bone mineral density: a population-based study [corrected] from the Canadian Multicentre Osteoporosis Study (CaMos). *Am. J. Gastroenterol.* 2012; 107: 1361–9.

[50] Fraser, L., Leslie, W. D., Targownik, L. E., Papaioannou, A., Adachi, J. D. The effect of proton pump inhibitors on fracture risk: report from the Canadian Multicenter Osteoporosis Study. *Osteoporos. Int.* 2013; 24: 1161–8.

[51] Lee, J., Youn, K., Choi, N.-K., Lee, J.-H., Kang, D., Song, H.-J., Park, B.-J. A population-based case-control study: proton pump inhibition and risk of hip fracture by use of bisphosphonate. *J. Gastroenterol.* 2013; 48: 1016–22.

[52] Cea Soriano, L., Ruigómez, A., Johansson, S., García Rodríguez, L. A. Study of the association between hip fracture and acid-suppressive drug use in a UK primary care setting. *Pharmacotherapy* 2014; 34: 570–81.

[53] Jo, Y., Park, E., Ahn, S. B., Jo, Y. K., Son, B., Kim, S. H., Sook, Y., Kim, H. J. A Proton Pump Inhibitor ' s Effect on Bone Metabolism Mediated by Osteoclast Action in Old Age : A Prospective Randomized Study. *Gut Liver* 2014; Epub. ahead.

[54] Nishi, T., Makuuchi, H., Weinstein, W. M. Changes in gastric ECL cells and parietal cells after long-term administration of high-dose omeprazole to patients with Barrett's esophagus. *Tokai J. Exp. Clin. Med.* 2005; 30: 117–121.

[55] Biosse-Duplan, M., Baroukh, B., Dy, M., de Vernejoul, M.-C., Saffar, J.-L. Histamine promotes osteoclastogenesis through the differential expression of histamine receptors on osteoclasts and osteoblasts. *Am. J. Pathol.* 2009; 174: 1426–1434.

[56] Abrahamsen, B., Vestergaard, P. Proton pump inhibitor use and fracture risk - effect modification by histamine H1 receptor blockade. Observational case-control study using National Prescription Data. *Bone* 2013; 57: 269–71.

[57] Abraham, N. S., Hlatky, M. A., Antman, E. M., Bhatt, D. L., Bjorkman, D. J., Clark, C. B., Furberg, C. D., Johnson, D. A., Kahi, C. J., Laine, L., Mahaffey, K. W., Quigley, E. M., Scheiman, J., Sperling, L. S., Tomaselli, G. F. ACCF/ACG/AHA 2010 expert consensus document on the concomitant use of proton pump inhibitors and thienopyridines: a focused update of the ACCF/ACG/AHA 2008 expert consensus

document on reducing the gastrointestinal risks of antiplatelet therapy and NSAID. *Am. J. Gastroenterol.* 2010; 105: 2533–49.

[58] Hamm, C. W., Bassand, J.-P., Agewall, S., Bax, J., Boersma, E., Bueno, H., Caso, P., Dudek, D., Gielen, S., Huber, K., Ohman, M., Petrie, M. C., Sonntag, F., Uva, M. S., Storey, R. F., Wijns, W., Zahger, D. ESC Guidelines for the management of acute coronary syndromes in patients presenting without persistent ST-segment elevation: The Task Force for the management of acute coronary syndromes (ACS) in patients presenting without persistent ST-segment elevatio. *Eur. Heart J.* 2011; 32: 2999–3054.

[59] Steg, P. G., James, S. K., Atar, D., Badano, L. P., Blömstrom-Lundqvist, C., Borger, M. A., Di Mario, C., Dickstein, K., Ducrocq, G., Fernandez-Aviles, F., Gershlick, A. H., Giannuzzi, P., Halvorsen, S., Huber, K., Juni, P., Kastrati, A., Knuuti, J., Lenzen, M. J., Mahaffey, K. W., Valgimigli, M., van 't Hof, A., Widimsky, P., Zahger, D. ESC Guidelines for the management of acute myocardial infarction in patients presenting with ST-segment elevation. *Eur. Heart J.* 2012; 33: 2569–619.

[60] Safety Labeling Information: Plavix [http://www.fda.gov/Safety/MedWatch/SafetyInformation/ucm225843.htm].

[61] Kazui, M., Nishiya, Y., Ishizuka, T., Hagihara, K., Farid, N. A., Okazaki, O., Ikeda, T., Kurihara, A. Identification of the human cytochrome P450 enzymes involved in the two oxidative steps in the bioactivation of clopidogrel to its pharmacologically active metabolite. *Drug Metab. Dispos.* 2010; 38: 92–99.

[62] Li, X. Q., Andersson, T. B., Ahlström, M., Weidolf, L. Comparison of inhibitory effects of the proton pump-inhibiting drugs omeprazole, esomeprazole, lansoprazole, pantoprazole, and rabeprazole on human cytochrome P450 activities. *Drug Metab. Dispos.* 2004; 32: 821–827.

[63] Ogilvie, B. W., Yerino, P., Kazmi, F., Buckley, D. B., Rostami-Hodjegan, A., Paris, B. L., Toren, P., Parkinson, A. The proton pump inhibitor, omeprazole, but not lansoprazole or pantoprazole, is a metabolism-dependent inhibitor of CYP2C19: Implications for coadministration with clopidogrel. *Drug Metab. Dispos.* 2011; 39: 2020–2033.

[64] Frelinger, A. L., Lee, R. D., Mulford, D. J., Wu, J., Nudurupati, S., Nigam, A., Brooks, J. K., Bhatt, D. L., Michelson, A. D. A randomized, 2-period, crossover design study to assess the effects of dexlansoprazole, lansoprazole, esomeprazole, and omeprazole on the steady-state

pharmacokinetics and pharmacodynamics of clopidogrel in healthy volunteers. *J. Am. Coll. Cardiol.* 2012; 59: 1304–1311.

[65] Parri, M. S., Gianetti, J., Dushpanova, A., Della Pina, F., Saracini, C., Marcucci, R., Giusti, B., Berti, S. Pantoprazole significantly interferes with antiplatelet effect of clopidogrel: Results of a pilot randomized trial. *Int. J. Cardiol.* 2013; 167: 2177–2181.

[66] Gilard, M., Arnaud, B., Cornily, J. C., Le Gal, G., Lacut, K., Le Calvez, G., Mansourati, J., Mottier, D., Abgrall, J. F., Boschat, J. Influence of Omeprazole on the Antiplatelet Action of Clopidogrel Associated With Aspirin. The Randomized, Double-Blind OCLA (Omeprazole CLopidogrel Aspirin) Study. *J. Am. Coll. Cardiol.* 2008; 51: 256–260.

[67] Kwok, C. S., Loke, Y. K. Effects of proton pump inhibitors on platelet function in patients receiving clopidogrel: a systematic review. *Drug Saf.* 2012; 35: 127–39.

[68] Hwang, S.-J., Jeong, Y.-H., Kim, I.-S., Koh, J.-S., Kang, M.-K., Park, Y., Kwak, C. H., Hwang, J.-Y. The cytochrome 2C19*2 and *3 alleles attenuate response to clopidogrel similarly in East Asian patients undergoing elective percutaneous coronary intervention. *Thromb. Res.* 2011; 127: 23–28.

[69] Hulot, J. S., Collet, J. P., Silvain, J., Pena, A., Bellemain-Appaix, A., Barthélémy, O., Cayla, G., Beygui, F., Montalescot, G. Cardiovascular Risk in Clopidogrel-Treated Patients According to Cytochrome P450 2C19*2 Loss-of-Function Allele or Proton Pump Inhibitor Coadministration. A Systematic Meta-Analysis. *J. Am. Coll. Cardiol.* 2010; 56: 134–143.

[70] Desta, Z., Zhao, X., Shin, J.-G., Flockhart, D. A. Clinical significance of the cytochrome P450 2C19 genetic polymorphism. *Clin. Pharmacokinet.* 2002; 41: 913–958.

[71] Sofi, F., Giusti, B., Marcucci, R., Gori, A. M., Abbate, R., Gensini, G. F. Cytochrome P450 2C19*2 polymorphism and cardiovascular recurrences in patients taking clopidogrel: a meta-analysis. *Pharmacogenomics J.* 2011; 11: 199–206.

[72] Ho, P. M., Maddox, T. M., Wang, L., Fihn, S. D., Jesse, R. L., Peterson, E. D., Rumsfeld, J. S. Risk of adverse outcomes associated with concomitant use of clopidogrel and proton pump inhibitors following acute coronary syndrome. *JAMA* 2009; 301: 937–944.

[73] Juurlink, D. N., Gomes, T., Ko, D. T., Szmitko, P. E., Austin, P. C., Tu, J. V., Henry, D. A., Kopp, A., Mamdani, M. M. A population-based

study of the drug interaction between proton pump inhibitors and clopidogrel. *CMAJ* 2009; 180: 713–8.

[74] Rassen, J. A., Choudhry, N. K., Avorn, J., Schneeweiss, S. Cardiovascular outcomes and mortality in patients using clopidogrel with proton pump inhibitors after percutaneous coronary intervention or acute coronary syndrome. *Circulation* 2009; 120: 2322–2329.

[75] Stockl, K. M., Le, L., Zakharyan, A., Harada, A. S. M., Solow, B. K., Addiego, J. E., Ramsey, S. Risk of rehospitalization for patients using clopidogrel with a proton pump inhibitor. *Arch. Intern. Med.* 2010; 170: 704–710.

[76] Van Boxel, O. S., van Oijen, M. G. H., Hagenaars, M. P., Smout, A. J. P. M., Siersema, P. D. Cardiovascular and gastrointestinal outcomes in clopidogrel users on proton pump inhibitors: results of a large Dutch cohort study. *Am. J. Gastroenterol.* 2010; 105: 2430–2436; quiz 2437.

[77] O'Donoghue, M. L., Braunwald, E., Antman, E. M., Murphy, S. A., Bates, E. R., Rozenman, Y., Michelson, A. D., Hautvast, R. W., Ver Lee, P. N., Close, S. L., Shen, L., Mega, J. L., Sabatine, M. S., Wiviott, S. D. Pharmacodynamic effect and clinical efficacy of clopidogrel and prasugrel with or without a proton-pump inhibitor: an analysis of two randomised trials. *Lancet* 2009; 374: 989–97.

[78] Bhatt, D. L., Cryer, B. L., Contant, C. F., Cohen, M., Lanas, A., Schnitzer, T. J., Shook, T. L., Lapuerta, P., Goldsmith, M. A., Laine, L., Scirica, B. M., Murphy, S. A., Cannon, C. P. Clopidogrel with or without omeprazole in coronary artery disease. *N. Engl. J. Med.* 2010; 363: 1909–1917.

[79] Roe, M. T., Armstrong, P. W., Fox, K. A. A., White, H. D., Prabhakaran, D., Goodman, S. G., Cornel, J. H., Bhatt, D. L., Clemmensen, P., Martinez, F., Ardissino, D., Nicolau, J. C., Boden, W. E., Gurbel, P. A., Ruzyllo, W., Dalby, A. J., McGuire, D. K., Leiva-Pons, J. L., Parkhomenko, A., Gottlieb, S., Topacio, G. O., Hamm, C., Pavlides, G., Goudev, A. R., Oto, A., Tseng, C.-D., Merkely, B., Gasparovic, V., Corbalan, R., Cinteză, M., et al. Prasugrel versus clopidogrel for acute coronary syndromes without revascularization. *N. Engl. J. Med.* 2012; 367: 1297–309.

[80] Siller-Matula, J. M., Jilma, B., Schrör, K., Christ, G., Huber, K. Effect of proton pump inhibitors on clinical outcome in patients treated with clopidogrel: a systematic review and meta-analysis. *J. Thromb. Haemost.* 2010; 8: 2624–2641.

[81] Focks, J., Brouwer, M., van Oijen, M., Lanas, A., Bhatt, D., Verheugt, F. Concomitant use of clopidogrel and proton pump inhibitors: impact on platelet function and clinical outcome- a systematic review. *Heart* 2013; 99: 520–527.

[82] Kwok, C. S., Loke, Y. K. Meta-analysis: the effects of proton pump inhibitors on cardiovascular events and mortality in patients receiving clopidogrel. *Aliment. Pharmacol. Ther.* 2010; 31: 810–823.

[83] Kwok, C. S., Jeevanantham, V., Dawn, B., Loke, Y. K. No consistent evidence of differential cardiovascular risk amongst proton-pump inhibitors when used with clopidogrel: Meta-analysis. *Int. J. Cardiol.* 2013; 167: 965–974.

[84] Douglas, I. J., Evans, S. J. W., Hingorani, A. D., Grosso, A. M., Timmis, A., Hemingway, H., Smeeth, L. Clopidogrel and interaction with proton pump inhibitors: comparison between cohort and within person study designs. *BMJ* 2012; 345: e4388.

[85] Wu, C. Y., Chan, F. K. L., Wu, M. S., Kuo, K. N., Wang, C. B., Tsao, C. R., Lin, J. T. Histamine 2-receptor antagonists are an alternative to proton pump inhibitor in patients receiving clopidogrel. *Gastroenterology* 2010; 139: 1165–1171.

[86] Chen, C. H., Yang, J. C., Uang, Y. S., Lin, C. J. Differential inhibitory effects of proton pump inhibitors on the metabolism and antiplatelet activities of clopidogrel and prasugrel. *Biopharm. Drug Dispos.* 2012; 33: 278–283.

[87] Theisen, J., Nehra, D., Citron, D. Suppression of gastric acid secretion in patients with gastroesophageal reflux disease results in gastric bacterial overgrowth and deconjugation of bile acids. *J. Gastrointest. Surg.* 2000; 4: 50–54.

[88] Thorens, J., Froehlich, F., Schwizer, W., Saraga, E., Bille, J., Gyr, K., Duroux, P., Nicolet, M., Pignatelli, B., Blum, A. L., Gonvers, J. J., Fried, M. Bacterial overgrowth during treatment with omeprazole compared with cimetidine: a prospective randomised double blind study. *Gut* 1996; 39: 54–59.

[89] Gulmez, S. E., Holm, A., Frederiksen, H., Jensen, T. G., Pedersen, C., Hallas, J. Use of proton pump inhibitors and the risk of community-acquired pneumonia: a population-based case-control study. *Arch. Intern. Med.* 2007; 167: 950–955.

[90] Sarkar, M., Hennessy, S., Yang, Y.-X. Proton-pump inhibitor use and the risk for community-acquired pneumonia. *Ann. Intern. Med.* 2008; 149: 391–398.

[91] Hermos, J. A., Young, M. M., Fonda, J. R., Gagnon, D. R., Fiore, L. D., Lawler, E. V. Risk of community-acquired pneumonia in veteran patients to whom proton pump inhibitors were dispensed. *Clin. Infect. Dis.* 2012; 54: 33–42.

[92] Johnstone, J., Nerenberg, K., Loeb, M. Meta-analysis: proton pump inhibitor use and the risk of community-acquired pneumonia. *Aliment. Pharmacol. Ther.* 2010; 31: 1165–1177.

[93] Giuliano, C., Wilhelm, S. M., Kale-Pradhan, P. B. Are proton pump inhibitors associated with the development of community-acquired pneumonia? A meta-analysis. *Expert Rev. Clin. Pharmacol.* 2012; 5: 337–44.

[94] Filion, K. B., Chateau, D., Targownik, L. E., Gershon, A., Durand, M., Tamim, H., Teare, G. F., Ravani, P., Ernst, P., Dormuth, C. R. Proton pump inhibitors and the risk of hospitalisation for community-acquired pneumonia: replicated cohort studies with meta-analysis. *Gut* 2014; 63: 552–8.

[95] Nardino, R. J., Vender, R. J., Herbert, P. N. Overuse of acid-suppressive therapy in hospitalized patients. *Am. J. Gastroenterol.* 2000; 95: 3118–3122.

[96] Herzig, S. J., Howell, M. D., Ngo, L. H., Marcantonio, E. R. Acid-suppressive medication use and the risk for hospital-acquired pneumonia. *JAMA* 2009; 301: 2120–2128.

[97] Jena, A., Sun, E., Goldman, D. Confounding in the association of proton pump inhibitor use with risk of community-acquired pneumonia. *J. Gen. Intern. Med.* 2012; 28: 223–30.

[98] De Groot, M. C. H., Klungel, O. H., Leufkens, H. G. M., van Dijk, L., Grobbee, D. E., van de Garde, E. M. W. Sources of heterogeneity in case-control studies on associations between statins, ACE-inhibitors, and proton pump inhibitors and risk of pneumonia. *Eur. J. Epidemiol.* 2014; 29: 767–75.

[99] Nerandzic, M. M., Pultz, M. J., Donskey, C. J. Examination of potential mechanisms to explain the association between proton pump inhibitors and Clostridium difficile infection. *Antimicrob. Agents Chemother.* 2009; 53: 4133–7.

[100] Antharam, V. C., Li, E. C., Ishmael, A., Sharma, A., Mai, V., Rand, K. H., Wang, G. P. Intestinal dysbiosis and depletion of butyrogenic bacteria in Clostridium difficile infection and nosocomial diarrhea. *J. Clin. Microbiol.* 2013; 51: 2884–2892.

[101] Chang, J. Y., Antonopoulos, D. A., Kalra, A., Tonelli, A., Khalife, W. T., Schmidt, T. M., Young, V. B. Decreased diversity of the fecal Microbiome in recurrent Clostridium difficile-associated diarrhea. *J. Infect. Dis.* 2008; 197: 435–438.

[102] Garcia-Mazcorro, J. F., Suchodolski, J. S., Jones, K. R., Clark-Price, S. C., Dowd, S. E., Minamoto, Y., Markel, M., Steiner, J. M., Dossin, O. Effect of the proton pump inhibitor omeprazole on the gastrointestinal bacterial microbiota of healthy dogs. *FEMS Microbiol. Ecol.* 2012; 80: 624–636.

[103] Seto, C. T., Jeraldo, P., Orenstein, R., Chia, N., DiBaise, J. K. Prolonged use of a proton pump inhibitor reduces microbial diversity: implications for Clostridium difficile susceptibility. *Microbiome* 2014; 2: 42.

[104] Leonard, J., Marshall, J. K., Moayyedi, P. Systematic review of the risk of enteric infection in patients taking acid suppression. *Am. J. Gastroenterol.* 2007; 102: 2047–2056.

[105] Janarthanan, S., Ditah, I., Adler, D. G., Ehrinpreis, M. N. Clostridium difficile-associated diarrhea and proton pump inhibitor therapy: a meta-analysis. *Am. J. Gastroenterol.* 2012; 107: 1001–10.

[106] Kwok, C. S., Arthur, A. K., Anibueze, C. I., Singh, S., Cavallazzi, R., Loke, Y. K. Risk of Clostridium difficile infection with acid suppressing drugs and antibiotics: meta-analysis. *Am. J. Gastroenterol.* 2012; 107: 1011–9.

[107] Deshpande, A., Pant, C., Pasupuleti, V., Rolston, D. D. K., Jain, A., Deshpande, N., Thota, P., Sferra, T. J., Hernandez, A. V. Association between proton pump inhibitor therapy and Clostridium difficile infection in a meta-analysis. *Clin. Gastroenterol. Hepatol.* 2012; 10: 225–33.

[108] Tleyjeh, I. M., Bin Abdulhak, A. A., Riaz, M., Alasmari, F. A., Garbati, M. A., AlGhamdi, M., Khan, A. R., Al Tannir, M., Erwin, P. J., Ibrahim, T., Allehibi, A., Baddour, L. M., Sutton, A. J. Association between proton pump inhibitor therapy and clostridium difficile infection: a contemporary systematic review and meta-analysis. *PLoS One* 2012; 7: e50836.

[109] Khanna, S., Aronson, S. L., Kammer, P. P., Baddour, L. M., Pardi, D. S. Gastric acid suppression and outcomes in Clostridium difficile infection: a population-based study. *Mayo Clin. Proc.* 2012; 87: 636–42.

[110] Buendgens, L., Bruensing, J., Matthes, M., Dückers, H., Luedde, T., Trautwein, C., Tacke, F., Koch, A. Administration of proton pump inhibitors in critically ill medical patients is associated with increased

risk of developing Clostridium difficile-associated diarrhea. *J. Crit. Care* 2014; 29: 696.e11–5.

[111] Freedberg, D. E., Salmasian, H., Friedman, C., Abrams, J. A. Proton pump inhibitors and risk for recurrent Clostridium difficile infection among inpatients. *Am. J. Gastroenterol.* 2013; 108: 1794–801.

[112] Cohen, S. H., Gerding, D. N., Johnson, S., Kelly, C. P., Loo, V. G., McDonald, L. C., Pepin, J., Wilcox, M. H. Clinical practice guidelines for Clostridium difficile infection in adults: 2010 update by the society for healthcare epidemiology of America (SHEA) and the infectious diseases society of America (IDSA). *Infect. Control Hosp. Epidemiol.* 2010; 31: 431–55.

[113] Surawicz, C. M., Brandt, L. J., Binion, D. G., Ananthakrishnan, A. N., Curry, S. R., Gilligan, P. H., McFarland, L. V., Mellow, M., Zuckerbraun, B. S. Guidelines for diagnosis, treatment, and prevention of Clostridium difficile infections. *Am. J. Gastroenterol.* 2013; 108: 478–98.

[114] Van Vlerken, L. G., Huisman, E. J., van Hoek, B., Renooij, W., de Rooij, F. W. M., Siersema, P. D., van Erpecum, K. J. Bacterial infections in cirrhosis: role of proton pump inhibitors and intestinal permeability. *Eur. J. Clin. Invest.* 2012; 42: 760–7.

[115] Chang, C. S., Chen, G. H., Lien, H. C., Yeh, H. Z. Small intestine dysmotility and bacterial overgrowth in cirrhotic patients with spontaneous bacterial peritonitis. *Hepatology* 1998; 28: 1187–1190.

[116] Parkman, H. P., Urbain, J. L., Knight, L. C., Brown, K. L., Trate, D. M., Miller, M. A., Maurer, A. H., Fisher, R. S. *Effect of Gastric Acid Suppressants on Human Gastric Motility. Volume 42*; 1998: 243–250.

[117] Parekh, P. J., Arusi, E., Vinik, A. I., Johnson, D. A. The role and influence of gut microbiota in pathogenesis and management of obesity and metabolic syndrome. *Front. Endocrinol. (Lausanne)* 2014; 5: 1–7.

[118] Lombardo, L., Foti, M., Ruggia, O., Chiecchio, A. Increased incidence of small intestinal bacterial overgrowth during proton pump inhibitor therapy. *Clin. Gastroenterol. Hepatol.* 2010; 8: 504–508.

[119] Compare, D., Pica, L., Rocco, A., De Giorgi, F., Cuomo, R., Sarnelli, G., Romano, M., Nardone, G. Effects of long-term PPI treatment on producing bowel symptoms and SIBO. *Eur. J. Clin. Invest.* 2011; 41: 380–386.

[120] Jacobs, C., Coss Adame, E., Attalure, A., Valestin, J., Rao, S. Dysmotility and PPI use are independent risk factors for small intestinal

bacterial and/or fungal overgrowth. *Aliment. Pharmacol. Ther.* 2013; 37: 1103–1111.

[121] Choung, R. S., Ruff, K. C., Malhotra, A., Herrick, L., Locke, G. R., Harmsen, W. S., Zinsmeister, A. R., Talley, N. J., Saito, Y. A. Clinical predictors of small intestinal bacterial overgrowth by duodenal aspirate culture. *Aliment. Pharmacol. Ther.* 2011; 33: 1059–1067.

[122] Ratuapli, S. K., Ellington, T. G., O'Neill, M.-T., Umar, S. B., Harris, L. A., Foxx-Orenstein, A. E., Burdick, G. E., Dibaise, J. K., Lacy, B. E., Crowell, M. D. Proton pump inhibitor therapy use does not predispose to small intestinal bacterial overgrowth. *Am. J. Gastroenterol.* 2012; 107: 730–5.

[123] Lo, W.-K., Chan, W. W. Proton pump inhibitor use and the risk of small intestinal bacterial overgrowth: a meta-analysis. *Clin. Gastroenterol. Hepatol.* 2013; 11: 483–90.

[124] Bajaj, J. S., Zadvornova, Y., Heuman, D. M., Hafeezullah, M., Hoffmann, R. G., Sanyal, A. J., Saeian, K. Association of proton pump inhibitor therapy with spontaneous bacterial peritonitis in cirrhotic patients with ascites. *Am. J. Gastroenterol.* 2009; 104: 1130–4.

[125] Choi, E. J., Lee, H. J., Kim, K. O., Lee, S. H., Eun, J. R., Jang, B. I., Kim, T. N. Association between acid suppressive therapy and spontaneous bacterial peritonitis in cirrhotic patients with ascites. *Scand. J. Gastroenterol.* 2011; 46: 616–620.

[126] Miura, K., Tanaka, A., Yamamoto, T., Adachi, M., Takikawa, H. Proton Pump Inhibitor Use is Associated with Spontaneous Bacterial Peritonitis in Patients with Liver Cirrhosis. *Intern. Med.* 2014; 53: 1037–1042.

[127] Goel, G. A., Deshpande, A., Lopez, R., Hall, G. S., van Duin, D., Carey, W. D. Increased rate of spontaneous bacterial peritonitis among cirrhotic patients receiving pharmacologic acid suppression. *Clin. Gastroenterol. Hepatol.* 2012; 10: 422–7.

[128] Trikudanathan, G., Israel, J., Cappa, J., O'Sullivan, D. M. Association between proton pump inhibitors and spontaneous bacterial peritonitis in cirrhotic patients - A systematic review and meta-analysis. *Int. J. Clin. Pract.* 2011; 65: 674–678.

[129] Yoshida, N., Yoshikawa, T., Tanaka, Y., Fujita, N., Kassai, K., Naito, Y., Kondo, M. A new mechanism for anti-inflammatory actions of proton pump inhibitors--inhibitory effects on neutrophil-endothelial cell interactions. *Aliment. Pharmacol. Ther.* 2000; 14 Suppl. 1: 74–81.

[130] Zedtwitz-Liebenstein, K., Wenisch, C., Patruta, S., Parschalk, B., Daxböck, F., Graninger, W. Omeprazole treatment diminishes intra- and

extracellular neutrophil reactive oxygen production and bactericidal activity. *Crit. Care Med.* 2002; 30: 1118–1122.

[131] Garcia-Tsao, G., Lee, F. A. Y., Barden, G. E., Cartun, R., West, A. B. Bacterial translocation to mesenteric lymph nodes is increased in cirrhotic rats with ascites. *Gastroenterology* 1995; 108: 1835–1841.

[132] Llovet, J. M., Bartoli, R., Planas, R., Cabre, E., Jimenez, M., Urban, A., Ojanguren, I., Arnal, J., Gassull, M. A. Bacterial translocation in cirrhotic rats. Its role in the development of spontaneous bacterial peritonitis. *Gut* 1994: 1648–1652.

[133] Campbell, M. S., Obstein, K., Reddy, K. R., Yang, Y.-X. Association between proton pump inhibitor use and spontaneous bacterial peritonitis. *Dig. Dis. Sci.* 2008; 53: 394–398.

[134] Miura, K., Ohnishi, H. Role of gut microbiota and Toll-like receptors in nonalcoholic fatty liver disease. *World J. Gastroenterol.* 2014; 20: 7381–91.

[135] Bulsiewicz, W. J., Scherer, J. R., Feinglass, J. M., Howden, C. W., Flamm, S. L. Proton Pump Inhibitor (PPI) Use Is Independently Associated with Spontaneous Bacterial Peritonitis (SBP) in Cirrhotics with Ascites. *Gastroenterology* 2009; 136: A–11.

[136] Thjodleifsson, B. Treatment of Acid-Related Diseases in the Elderly with Emphasis on the Use of Proton Pump Inhibitors. *Drugs Aging* 2002; 19: 911–927.

[137] Kwon, J. H., Koh, S.-J., Kim, W., Jung, Y. J., Kim, J. W., Kim, B. G., Lee, K. L., Im, J. P., Kim, Y. J., Kim, J. S., Yoon, J.-H., Lee, H.-S., Jung, H. C. Mortality associated with proton pump inhibitors in cirrhotic patients with spontaneous bacterial peritonitis. *J. Gastroenterol. Hepatol.* 2014; 29: 775–81.

[138] Min, Y. W., Lim, K. S., Min, B.-H., Gwak, G.-Y., Paik, Y. H., Choi, M. S., Lee, J. H., Kim, J. J., Koh, K. C., Paik, S. W., Yoo, B. C., Rhee, P.-L. Proton pump inhibitor use significantly increases the risk of spontaneous bacterial peritonitis in 1965 patients with cirrhosis and ascites: a propensity score matched cohort study. *Aliment. Pharmacol. Ther.* 2014; 40: 695–704.

[139] Dultz, G., Piiper, A., Zeuzem, S., Kronenberger, B., Waidmann, O. Proton pump inhibitor treatment is associated with the severity of liver disease and increased mortality in patients with cirrhosis. *Aliment. Pharmacol. Ther.* 2014.

[140] Dial, M. S. Proton pump inhibitor use and enteric infections. *Am. J. Gastroenterol.* 2009; 104 Suppl.: S10–S16.

[141] Cobelens, F. G., Leentvaar-Kuijpers, A., Kleijnen, J., Coutinho, R. A. Incidence and risk factors of diarrhoea in Dutch travellers: consequences for priorities in pre-travel health advice. *Trop. Med. Int. Health* 1998; 3: 896–903.

[142] Bavishi, C., Dupont, H. L. Systematic review: the use of proton pump inhibitors and increased susceptibility to enteric infection. *Aliment. Pharmacol. Ther.* 2011; 34: 1269–81.

[143] Giannella, R. A., Broitman, S. A., Zamcheck, N. Gastric acid barrier to ingested microorganisms in man: studies in vivo and in vitro. *Gut* 1972; 13: 251–256.

[144] Tennant, S. M., Hartland, E. L., Phumoonna, T., Lyras, D., Rood, J. I., Robins-Browne, R. M., Van Driel, I. R. Influence of gastric acid on susceptibility to infection with ingested bacterial pathogens. *Infect. Immun.* 2008; 76: 639–645.

[145] Lin, J., In Soo Lee, Frey, J., Slonczewski, J. L., Foster, J. W. Comparative analysis of extreme acid survival in Salmonella typhimurium, Shigella flexneri, and Escherichia coli. *J. Bacteriol.* 1995; 177: 4097–4104.

[146] Foster, J. W. Escherichia coli acid resistance: tales of an amateur acidophile. *Nat. Rev. Microbiol.* 2004; 2: 898–907.

[147] DuPont, H., Formal, S., Hornick, R., Et, A. Pathogenesis of Escherichia coli diarrhea. *N. Engl. J. Med.* 1971; 285: 1–9.

[148] Dupont, H. L., Ericsson, C. D., Farthing, M. J. G., Gorbach, S., Pickering, L. K., Rombo, L., Steffen, R., Weinke, T. Expert review of the evidence base for prevention of travelers' diarrhea. *J. Travel Med.* 2009; 16: 149–160.

[149] Kodner, C. M., Kudrimoti, A. Diagnosis and Management of Acute Interstitial Nephritis. *Am. Fam. Physician* 2003; 67: 2527–2534.

[150] Leonard, C. E., Freeman, C. P., Newcomb, C. W., Reese, P. P., Herlim, M., Bilker, W. B., Hennessy, S., Strom, B. L. Proton pump inhibitors and traditional nonsteroidal anti-in flammatory drugs and the risk of acute interstitial nephritis and acute kidney injury. *Pharmacoepidemiol. Drug Saf.* 2012; 21: 1155–1172.

[151] Sampathkumar, K., Ramalingam, R., Prabakar, A., Abraham, A. Acute interstitial nephritis due to proton pump inhibitors. *Indian J. Nephrol.* 2013; 23: 304–7.

[152] Sierra, F., Suarez, M., Rey, M., Vela, M. F. Systematic review: Proton pump inhibitor-associated acute interstitial nephritis. *Aliment. Pharmacol. Ther.* 2007; 26: 545–553.

[153] Methotrexate Safety information [http://www.fda.gov/Drugs/InformationOnDrugs/ApprovedDrugs/ucm284421.htm].

[154] Breedveld, P., Zelcer, N., Pluim, D., Tibben, M. M., Beijnen, J. H., Schinkel, A. H., Tellingen, O. Van, Borst, P., Schellens, J. H. M. Mechanism of the Pharmacokinetic Interaction between Methotrexate and Benzimidazoles: Potential Role for Breast Cancer Resistance Protein in Clinical Drug-Drug Interactions. *Cancer Res.* 2004; 64: 5804–5811.

[155] Joerger, M., Huitema, A. D. R., Van Den Bongard, H. J. G. D., Baas, P., Schornagel, J. H., Schellens, J. H. M., Beijnen, J. H. Determinants of the elimination of methotrexate and 7-hydroxy-methotrexate following high-dose infusional therapy to cancer patients. *Br. J. Clin. Pharmacol.* 2006; 62: 71–80.

[156] Suzuki, K., Doki, K., Homma, M., Tamaki, H., Hori, S., Ohtani, H., Sawada, Y., Kohda, Y. Co-administration of proton pump inhibitors delays elimination of plasma methotrexate in high-dose methotrexate therapy. *Br. J. Clin. Pharmacol.* 2009; 67: 44–49.

[157] Reeves, D. J., Moore, E. S., Bascom, D., Rensing, B. Retrospective evaluation of methotrexate elimination when co-administered with proton pump inhibitors. *Br. J. Clin. Pharmacol.* 2014; 78: 565–71.

[158] Bezabeh, S., Mackey, A. C., Kluetz, P., Jappar, D., Korvick, J. Accumulating evidence for a drug-drug interaction between methotrexate and proton pump inhibitors. *Oncologist* 2012; 17: 550–4.

[159] Wiviott, S. D., Trenk, D., Frelinger, A. L., O'Donoghue, M., Neumann, F.-J., Michelson, A. D., Angiolillo, D. J., Hod, H., Montalescot, G., Miller, D. L., Jakubowski, J. A., Cairns, R., Murphy, S. A., McCabe, C. H., Antman, E. M., Braunwald, E. Prasugrel compared with high loading- and maintenance-dose clopidogrel in patients with planned percutaneous coronary intervention: the Prasugrel in Comparison to Clopidogrel for Inhibition of Platelet Activation and Aggregation-Thrombolysis in Myocardial. *Circulation* 2007; 116: 2923–32.

[160] Wiviott, S. D., Braunwald, E., McCabe, C. H., Montalescot, G., Ruzyllo, W., Gottlieb, S., Neumann, F.-J., Ardissino, D., De Servi, S., Murphy, S. A., Riesmeyer, J., Weerakkody, G., Gibson, C. M., Antman, E. M. Prasugrel versus clopidogrel in patients with acute coronary syndromes. *N. Engl. J. Med.* 2007; 357: 2001–2015.

In: Proton Pump Inhibitors (PPIs)
Editor: Barbara Parker
ISBN: 978-1-63482-890-1

Chapter V

Adverse Events of Long-Term Proton Pump Inhibitor Therapy[*]

Nieka K. Harris, MD, Amy N. Stratton, DO, and Fouad J. Moawad[†], MD
Department of Medicine, Gastroenterology Service, Walter Reed National Military Medical Center, Bethesda, MD, US

Abstract

Gastroesophageal reflux disease (GERD) is a common condition affecting approximately 10-30% of the Western population. Risk factors include obesity, diet, anatomic factors such as a hiatal hernia, and medications which may lower the lower esophageal sphincter (LES) resting pressure. With the rise in risk factors and subsequent increase in gastroesophageal reflux symptoms, proton pump inhibitors (PPI) are commonly prescribed and used long-term. Although PPIs are relatively safe and necessary when indicated, there are concerns regarding long-term adverse effects of chronic PPI use. With decreased gastric acid

---

[*] The views expressed in this abstract/manuscript are those of the author(s) and do not reflect the official policy or position of the Department of the Army, Department of the Navy, Department of Defense, or the U.S. Government.

[†] Corresponding Author: Walter Reed National Military Medical Center. 8901 Wisconsin Ave., Bethesda, MD 20889, E-mail: fouad.j.moawad.mil@mail.mil. Phone: 301-295-5035. Fax: 301-295-459.

secretion, there exists a potential for micronutrient malabsorption, osteoporosis, and enteric infections. Less common complications of PPI usage include community associated pneumonia, interstitial nephritis and microscopic colitis. In this review, we explore the current literature on potential complications associated with long-term PPI use.

Funding sources: Grant support: None
Disclosures: Conflicts of interest: None

Introduction

Proton pump inhibitors (PPIs) are potent antagonists of gastric acid production which are very effective in treating upper gastrointestinal tract acid-mediated disease conditions. PPIs are commonly prescribed in the United States with annual sales exceeding $11 billion [1]. They are routinely used in the treatment of symptomatic gastroesophageal reflux disease (GERD), peptic ulcer disease (PUD), non-ulcer dyspepsia, and *Helicobacter pylori* eradication. The developing obesity epidemic and resultant increase in gastroesophageal reflux symptoms has led to an increase in long-term PPI use. Self-prescribed PPI use is also on the rise with easy access to over the counter (OTC) omeprazole (Prilosec® OTC), esomeprazole (Nexium® 24HR), and lansoprazole (Prevacid® 24HR).

In 1976, the gastric H^+/K^+/ATPase, known as the proton pump, was first discovered [2]. Shortly thereafter timoprazole was developed and found to be a potent inhibitor of gastric acid secretion [2]. However, it was not until 1989 that omeprazole gained approval by the FDA as the first PPI on the market. This resulted in a shift in the medical management of GERD and other acid hypersecretory disorders. Currently, there are six PPIs available for use and include dexlansoprazole, esomeprazole, lansoprazole, omeprazole, pantoprazole, and rabeprazole.

Common adverse side effects associated with PPIs are nausea, abdominal pain, flatulence, diarrhea, and headache [3]. Yet, with growing experience in long-term use of PPIs there are increasing reports of more concerning adverse events. Epidemiological studies have linked PPIs to several nutritional deficiencies, metabolic and infectious diseases. Other reported complications include microscopic colitis and interstitial nephritis. The aim of this article is to further explore the adverse effects associated with long-term PPI use.

Rebound Acid Hypersecretion

With acid suppression, elevated gastric pH stimulates gastrin release. This, in turn, stimulates enterochromaffin-like cell (ECL) to increase acid secretion in the stomach [4]. With continued proton pump inhibition, these cells remain suppressed; however, abrupt cessation of a PPI can lead to unregulated hypersecretion of gastric acid. A randomized placebo controlled trial in 119 healthy subjects demonstrated that 44% of subjects treated with a PPI for 8 weeks compared to 7% of subjects treated with placebo developed reflux symptoms upon withdrawal of drug [5]. This was confirmed in a Swedish study in which 44% of healthy subjects developed reflux symptoms with cessation of a PPI after 4 weeks of treatment compared to 9% of subjects receiving placebo [4]. Three other studies have demonstrated evidence of clinically symptomatic rebound acid hypersecretion, even after only 5 days of PPI treatment [5]. Thus, development or recurrence of GER symptoms is a potential complication of abrupt PPI cessation and patients should be counseled regarding this risk. To reduce symptoms associated with acid hypersecretion, consideration should be given to either a slow PPI taper or on-demand use of H2 receptor antagonists (H2RAs) for 7-14 days during withdrawal.

Micronutrient Malabsorption

There have been reports of multiple micronutrient deficiencies associated with long-term PPI use, which include malabsorption of magnesium, calcium, iron, and vitamin B12. These nutrient deficiencies are believed to be secondary to the hypochlorhydria resulting from chronic PPI use. Of these minerals and vitamins, the association with hypomagnesaemia is the strongest. On the other hand, the association between long-term PPI use and iron or vitamin B12 deficiency has not been well established.

Hypomagnesemia

There are two mechanisms of intestinal absorption of magnesium. One involves passive absorption without the use of membrane channels, whereas the other involves co-transport of magnesium and calcium via the TRPM6 and

TRPM7 channels. TRPM6/ TRPM7 channel activity has been shown to increase in an acid rich environment in both animal and human models [6-8]. Therefore, with acid suppression secondary to PPI, reduced TRPM6/TRPM7 activity occurs followed by magnesium malabsorption. In one study, omeprazole selectively inhibited the passive intestinal absorption of magnesium, supporting a separate mechanism of decreased absorption [9].

There have been several case reports of hypomagnesemia in patients taking long-term PPIs. The majority of these cases were asymptomatic and occurred with at least 1 year of use, and was especially more prevalent with greater than 5 years of use [3, 4]. A small subset of patients developed symptomatic hypomagnesemia characterized by ataxia, paresthesias, tetany, seizures, arrhythmias or tremors requiring hospitalization [4,10]. On follow up evaluation, the hypomagnesemia resolved in these reported cases upon cessation of PPI therapy. To establish causality, a portion of the reported cases was re-challenged with PPI therapy and subsequently redeveloped hypomagnesemia with resolution on withdrawal of the PPI [3, 10].

Hypocalcemia and Risk of Fractures

In order to be intestinally absorbed, dietary calcium requires ionization in an acid- rich environment [3]. Hypochlorhydria, secondary to chronic PPI use, may lead to decreased calcium ionization and absorption. Multiple small crossover and randomized controlled studies have demonstrated mixed results regarding changes in calcium absorption in relation to PPI therapy [11]. In one study, fractional calcium absorption (FCA) was evaluated in 23 postmenopausal women with no history of use of acid suppressants, medication use known to affect calcium absorption, or medical conditions associated with malabsorption [12]. Subjects completed two 24-hour calcium absorption studies to include one after 30 days of omeprazole treatment daily [12]. There was no evidence of calcium malabsorption with short-term PPI ingestion with a baseline FCA of 18-20% and post PPI exposure FCA of 23% ($p = 0.07$).

Acid suppression with PPIs may impact fracture risk. A number of retrospective studies evaluating a link between PPI use and decreased bone mineral density (BMD) have yielded conflicting results. Additionally, omeprazole is reported to decrease osteoclastic activity which may lead to an inability to repair micro-fractures resulting in an increase in fracture risk [10, 13]. A meta-analysis of observational studies evaluating the risk of fracture with PPI use in 223,210 patients found a pooled odds ratios (ORs) for hip fracture of 1.25, vertebral fracture of 1.5, and forearm/wrist fracture of 1.09 in patients taking PPI [14]. Interestingly, short duration of PPI use was associated with an increased hip fracture risk, whereas long-term use was not [14]. BMD in relation to PPI and H2RA use was also assessed in the longitudinal observational Study of Women Across the Nation (SWAN) cohort [15]. The study evaluated a total of 2,068 peri-menopausal women who were either taking PPI, H2RA or were non-users. Estimated annual spine BMD loss was -0.58 (95% CI: -0.62 to -0.52) in non-users compared to -0.51 (95% CI: -0.65 to -0.37) in H2RA users and -0.48 (95% CI: -0.62 to -0.33) in PPI users [15]. Similar patterns of estimated annual BMD loss were seen in the femoral neck and hip [15].

In 2010, the Food and Drug Administration (FDA) issued a warning regarding a potential increased fracture risk associated with PPI use and as such patients should be counseled appropriately [10]. Some studies have reported that the combined use of PPI and bisphosphonate leads to an increase fracture risk through attenuation of bisphosphonate activity [16]. Special caution should be used in prescribing short and long-term PPIs to patients older than 65 years of age, other risk factors for bone disease, or those with documented osteopenia or osteoporosis.

Vitamin B12 Deficiency

Vitamin B12, also known as cobalamin, is a water-soluble vitamin that is bound to dietary protein. In the acidic environment of the stomach, hydrogen ions and pepsin release B12 from dietary protein. B12 then binds to salivary protein facilitating transfer to intrinsic factor where the complex can later be absorbed in the terminal ileum. In the absence of hydrogen ion secretion, B12 may not be efficiently released from dietary protein ultimately leading to decreased absorption.

Valuck et al. performed a retrospective case-control study of 53 patients who were 75 years or older with vitamin B12 deficiency compared to 212 controls [17]. Past or short-term current use, defined as less than 12 months of PPI or H2RA, was not associated with increased risk of B12 deficiency (OR 2.01; 95% CI: 0.89 - 4.35 and OR 1.03; 95% CI: 0.46 - 2.31, respectively) [17]. However, the study demonstrated an increased risk of vitamin B12 deficiency with current use of PPI or H2RA for 12 months or longer (OR 4.46; 95% CI: 1.49 - 13.33, respectively) [17]. A Dutch prospective cross-sectional study of long-term PPI use, defined as greater than 3 years, in 125 couples age 65 or older showed no increase in B12 deficiency with PPI use [18]. Of those patients taking a PPI, 3% were diagnosed with B12 deficiency compared to 2% of their partners who were not taking any form of acid suppressant ($p = 1.00$) [18].

Lam et al. performed a nested case-control study evaluating 25,956 subjects with B12 deficiency concurrently prescribed 2 or more years of PPI to 184,199 matched controls. Subjects taking PPIs for greater than 2 years were more likely to have incident diagnoses of vitamin B12 deficiency compared to matched controls (OR 1.65; 95% CI: 1.58 - 1.73) [19]. There also appears to be a dose and duration dependent increase in vitamin B12 deficiency with PPI use [4,19]. However, this study did not assess symptoms of B12 deficiency between the two cohorts, and thus the clinical significance of relative B12 deficiency in the setting of PPI use is not known.

Although the current body of literature suggests a potential association between vitamin B12 deficiency and PPI use, data is primarily derived from retrospective studies with confounding factors which limits their conclusions, and therefore routine monitoring of vitamin B12 levels in patients taking longer term PPIs is not currently recommended.

Iron Deficiency

Non-heme iron is ingested as ferric ions and requires a pH less ~~of~~ than 3 to be reduced to the more soluble ferrous state [3]. A study evaluated radiolabeled iron absorption in healthy individuals taking cimetidine and found a 28-65% dose dependent reduction in iron absorption [3,20].

In a retrospective cohort study of 98 PPI users compared to an equal number of age and sex matched controls, PPI users were more likely to have a $\leq$1 g/dL decrease in their hemoglobin (OR 5.03; 95% CI: 1.71 - 14.78), however a decrease in mean corpuscular volume (MCV) was not significant (OR 1.77; 95% CI: 0.77 - 4.05) [21]. This study demonstrated a potential association between PPI use and anemia, although not a statistically significant decrease in the MCV, which would be expected in iron deficiency.

Data is sparse with regards to PPI use and iron deficiency. Hutchinson et al. observed decreased phlebotomy requirements in 7 hemochromatosis patients taking PPI to maintain a ferritin of 50 μg/L. Patients were also observed to have decreased non-heme iron absorption from a meal after taking a PPI for 7 days [22]. However, a study of 109 subjects with Zollinger-Ellison syndrome taking omeprazole for greater than 6 years failed to demonstrate evidence of iron deficiency [23]. The association between iron deficiency and long-term PPI use is biologically plausible, but has not been firmly established and as such routine monitoring of iron and ferritin levels are not recommended in patients taking PPIs.

Infections

Clostridium Difficile Infection (CDI)

A prospective cohort study in 101,796 patients over a five year period found an increased risk of CDI with increasing levels of acid suppression from an OR of 1 with no acid suppression to 1.53 for H2RA use only, to 1.74 for daily PPI use, and 2.36 for PPI use greater than once daily [24]. In a meta-analysis of 30 studies including over 202,000 patients, the risk of CDI was greater in patients exposed to PPI therapy (OR 2.15; 95% CI: 1.81 - 2.55) [25]. A subsequent meta-analysis including 39 studies, 30 case-control and 12 cohort, with a total of 313,000 subjects noted a 1.74 fold increased risk for CDI in PPI users [26]. In a third meta-analysis of 15 case-control and 6 cohort studies with greater than 280,000 subjects, a greater than 60% increase in CDI incidence was noted among patients on PPIs with a relative risk (RR) of 1.69 [27]. The collective results of these studies have led the FDA to release a warning that PPIs may increase the risk of CDI.

There are several potential reasons as to why PPIs increase CDI risk. The intestinal microbiota play an important function in suppression of pathogen growth and alterations in the microbiota by PPIs may support pathogenic *C. difficile* proliferation [28]. Microbiome analysis of the feces has been shown to improve the ability to distinguish CDI status when comparing cases with diarrheal and non-diarrheal controls [28]. Prolonged use of PPIs has been linked to changes in the microbial community composition of the upper GI tract in vitro. Also, gastric acid reduction influences the microbial composition of the lower GI tract. The changes in the gut microbial ecology during and after PPI use and the long-term effect of PPI use relative to the microbiota present in CDI are not entirely clear, but there is evidence to support a potential increased risk of CDI in patients taking PPI.

Enteric Infections (Enterocolitis)

Gastric acid is a primary defense mechanism against ingested pathogens and thus neutralization with acid suppressive medications could lead to proliferation of acid sensitive pathogens. Studies have reported increased gastric bacterial concentrations in patients on PPI therapy [29]. *Camphylobacter* and *Salmonella* are acid-sensitive bacteria whose enteric infection rates have increased in patients on long-term PPIs. Multiple case-control studies have noted increased risk of *C. jejuni* with an OR between 1.7 and 11.7 [30, 31]. In a recent meta-analysis of studies evaluating *Salmonella*, *C. jejuni*, and other enteric infections in over 11,000 patients, an increase in enteric infections was noted [32]. The association was also greater in patients on PPIs compared to those taking H2RAs (OR 2.03; 95% CI: 1.05 -3.92) [32]. Risk of enteric infection with *Shigella* and *Listeria* monocytogenes does not increase in the setting of acid suppression, as they are acid-resistant pathogens [33, 34].

Parasitic infections are also of concern, with *Giardiasis* and *Strongliodiasis* being most frequently noted in the literature [35]. *Giardiasis* is the most common protozoal infection of the small intestine worldwide; however, gastric infections are rare. As *Giardia* is acid sensitive, there is a theoretical risk of gastric survival with hypochlorhydria. Cook noted more than 50% of intestinal *giardiasis* was associated with hypochlorhydria [36]. *Strongloides* hyper-infection has also been reported following H2RA use in immunosuppressed patients, but there is no known association in PPI users [37].

Spontaneous Bacterial Peritonitis (SBP)

In cirrhotic patients with ascites, SBP occurs secondary to bacterial translocation across the intestinal wall into the mesenteric lymph nodes. It is associated with decreased bacterial clearance, increased bacterial overgrowth, and altered motility in the setting of increased permeability. PPIs are noted to increase enteric bacterial colonization and overgrowth, which may increase the risk of SBP in the cirrhotic population on PPI therapy. Campbell et al. conducted a retrospective case-control study of 116 consecutive cirrhotic patients with ascites in which 32 patients developed SBP [38]. The OR for the development of SBP among PPI users versus non-users was 1.22 (95% CI: 0.52 - 2.87) [38]. Other studies have noted an increased risk of SBP among PPI users. A meta-analysis of eight observational studies investigating the association of SBP with acid suppressants in 3,815 patients noted a 3-fold increase in SBP (OR 3.15; 95% CI: 2.09-4.74) in patients on PPI versus 562 patients on H2RAs (OR 1.71; 95% CI: 0.97 -3.01) [39].

Small Intestinal Bacterial Overgrowth (SIBO)

SIBO occurs when the microbiota of the upper gastrointestinal tract is altered. Over the last decade, there has been mixed data associating SIBO with PPI use. The lack of diagnostic accuracy for SIBO makes this association even more challenging to establish. Lo et al. conducted a meta-analysis demonstrating a statistically significant association between PPI use and SIBO; however, this was only noted in the sub-group analysis in which SIBO was diagnosed via duodenal and jejunal aspirate cultures (OR 2.28; 95% CI: 1.24 - 4.21) [40]. Lombardo et al. evaluated 450 patients undergoing glucose hydrogen breath tests (GHBTs) in 200 patients with GERD on PPI for a median of 3 years, 200 patients with IBS in the absence of PPI use for 3 years, and 50 healthy controls not receiving PPIs for the past 10 years. SIBO was noted in 50% of the GERD cohort. Furthermore, an increased in incidence of SIBO was noted after only one year of PPI use [41].

A study of 42 patients with non-erosive reflux disease treated with twice daily PPI for 6 months found an increase in bowel symptoms (abdominal pain, bloating, diarrhea) and a positive GHBT in 26% of the patients [42]. In another retrospective study aimed to determine predictors of SIBO using duodenal aspirate culture, PPI use was not associated with SIBO [43].

In a retrospective chart review of 1,191 patients who underwent GHBT, of which nearly half the cohort was taking PPI, no significant association was seen between PPI and GHBT positivity [44]. Overall, there are no well controlled studies to firmly establish a cause and effect relationship between PPI and SIBO.

Community Acquired Pneumonia (CAP)

CAP is a serious clinical illness with significant morbidity and mortality. The increased gastric pH caused by acid suppressive therapy can theoretically permit pathogen survival and colonization of the upper GI tract. Micro-aspiration of the resultant bacteria into the lungs could then lead to development of CAP and bacterial translocation.

Randomized controlled trials have not provided a definitive conclusion regarding this association. Sultan et al. demonstrated no significant association between PPI use and respiratory infections in a meta-analysis of randomized controlled trials (OR 1.42; 95% CI: 0.86 - 2.35) [45]. Additionally, Estborn et al. found no causal association between esomeprazole use and CAP in a pooled analysis of 131 clinical trials with 16,583 patients receiving PPI [46]. In contrast, a nested case control study comparing the incidence of CAP in acid suppression medication users versus non-users found a 2 fold increase in the development of streptococcal pneumonia among PPI users [47].

Data from 6 observational studies pooled in a meta-analysis revealed that the overall risk for CAP was higher among PPI users (OR 1.36; 95% CI: 1.12 - 1.65) [48]. Subgroup analysis indicated that short duration of use (<30 days) was associated with increased odds of CAP (OR 1.92; 95% CI: 1.40 - 2.63), however chronic use was not (OR 1.11; 95% CI: 0.90 - 1.38) [48].

A meta-analysis of 9 case-controlled and cohort studies with 120,615 pneumonia cases demonstrated a significant association with PPI use (OR 1.39; 95% CI: 1.09-1.76) [49]. PPI use for less than 30 days (OR 1.65; 95% CI: 1.25 - 2.19), high dose PPI use (OR 1.50; 95% CI: 1.33 - 1.68), and low dose PPI (OR 1.17; 95% CI: 1.11 - 1.24) were all also significantly associated with CAP [49]. However, chronic use of PPI (>180 days) was not associated with CAP (OR 1.10; 95% CI: 1.00 - 1.21) [49]. In the pediatric literature, there is some evidence of pneumonia associated with PPI use in children [50].

The association between short-term PPI use and CAP seems biologically plausible. However, heterogeneity between studies and confounding factors limit some of the conclusions regarding the relationship between PPI and pneumonia.

Microscopic Colitis

Microscopic colitis is characterized by chronic watery diarrhea in the absence of infectious etiologies. Typically, the macroscopic appearance of the colonic mucosa is normal and the diagnosis is made microscopically. There are two categories of microscopic colitis: collagenous colitis (thickened subepithelial collagen layer) and lymphocytic colitis (increased surface epithelium lymphocytes) [51]. The proposed mechanisms involves either a genetic predisposition via TRPM6/TRPM7, conformational changes in cellular structure leading to increased cell membrane permeability, or hypochlorhydria resulting in changes in the intestinal microbiota with an increased risk for gastrointestinal infections [52].

A retrospective case-control study of 136 patients who received a PPI prescription within the prior 180 days were compared to 355 randomly age and gender matched patients from the general population [53]. Patients receiving

PPIs were more likely to develop microscopic colitis compared to those not exposed to PPIs (38% versus 13%, p <0.001) [53]. Of those patients with PPI-associated microscopic colitis, 40% were exposed to omeprazole, 23% to esomeprazole, 29% to pantoprazole, 6% to rabeprazole, and 3% to lansoprazole [53]. In a case-series of 8 subjects, the authors concluded that each subject had probable to definite lansoprazole-related microscopic colitis as determined by a causality score [51, 54]. In their literature review, 25 additional cases of purported lansoprazole related microscopic colitis were also found [51]. Keszthelyi et al. conducted a meta-analysis which supported a likely causal relationship between PPIs and microscopic colitis [52].

Acute Interstitial Nephritis (AIN)

AIN is a class of acute kidney injury consisting of an increase in serum creatinine associated with urine eosinophils and constitutional symptoms, rash, and peripheral eosinophilia [3,55]. The first case of PPI-associated AIN was reported by Ruffenach in 1992, secondary to omeprazole [56]. Muriithi et al. performed a retrospective case series of 133 subjects with biopsy-proven AIN and found that drug induced AIN was responsible for 70% of the 133 reported cases [55]. Of these cases, antibiotics and PPIs, most notably omeprazole, were the most common instigators, associated with 35% and 10% of all cases respectively [55]. The nephropathy appeared reversible as the majority of subjects had partial to complete recovery of kidney function 6 months after PPI cessation [55]. Similar to other cases of drug induced adverse events, initial treatment of AIN is removal of the culprit agent. Studies have shown that a decrease time to withdrawal of the offending medication leads to improved outcomes and recovery of renal function. There is conflicting evidence regarding the use of steroid therapy for treatment of drug induced AIN. While some studies report improved outcomes with the use of steroids, other studies have not demonstrated improvement in outcomes or recovery of renal function [3, 55, 56].

Medication Interactions

PPIs reduce gastric acid and subsequently affect the bioavailability of various drugs through competitive hepatic metabolism of the cytochrome P450 system, which is predominantly responsible for PPI metabolism. Most notably CYP 2C19, CYP 3A4 and CYP 11A2 are components of the enzyme system inhibited by PPIs [57]. Additionally, polymorphic genetic alleles of CYP2C19*2*3, which are found in various ethnic groups, should be considered as a contributing factor to PPI metabolism and its effects on the bioavailability of drugs metabolized by the cytochrome P450 system [58]. Examples of drugs associated with decreased bioavailability include ketoconazole, itraconazole, thyroxine, atazanavir, and dipyridamole, while other drugs such as nifedipine, digoxin, and alendronate are associated with increased PPI bioavailability [59-62].

Clopidogrel along with aspirin is considered standard of care for flow limiting coronary artery disease following drug eluting stent placement. Concomitant treatment with PPIs in order to prevent gastrointestinal side effects is currently recommended by consensus guidelines. Clopidogrel, a hepatically activated prodrug, requires metabolic activation by the same hepatic cytochrome P450 isoenzymes as PPI (CYP 2C19 and CYP 3A4). As such, concerns exist regarding the possible interaction between PPIs and clopidogrel that could reduce the antiplatelet and cardioprotective effects of clopidogrel.

Gilard et al. demonstrated in vitro reduction of the anti-aggregating activity of clopidogrel in patients after coronary revascularization under PPI treatment [63]. Two years later, the same group conducted a randomized double-blind placebo-controlled trial of 124 patients undergoing elective coronary artery stent placement. Patients receiving aspirin and clopidogrel were randomized to either omeprazole 20 mg daily or placebo [64]. The mean platelet reactivity index was higher in the omeprazole group (51.4% versus 39.8%, $p<0.001$, indicating less effective platelet anti-aggregation effect [64].

Based on Canadian and United States insurance records, Rassen et al. analyzed 18,565 patients hospitalized for ACS who underwent percutaneous intervention (PCI) [65]. Patient using clopidogrel and PPI had a slightly higher risk for re-hospitalization for myocardial infarction or death of any cause (RR 1.26; 95% CI: 0.97 - 1.63) [65]. A similar analysis was performed by Ray et al. evaluating 20,286 patients after hospitalization for ACS and PCI. However, concomitant clopidogrel and PPI use was not associated with serious cardiovascular disease (HR 0.99; 95% CI: 0.82 - 1.19) [66]. Sub-group analysis evaluating the different types of PPI did not identify any increased risk of serious cardiovascular disease [66].

Post-hoc analysis of the randomized Clopidogrel for Reduction of Events During Observation (CREDO) trial, Dunn et al. reported an increased risk of death, myocardial re-infarction or urgent target vessel revascularization at 28 days in patients using PPIs, independent of the underlying treatment [clopidogrel (OR 1.63; 95% CI: 1.02 - 2.63) or placebo (OR 1.55; 95% CI: 1.03 - 2.34)] [67].

In another randomized controlled double-blind multicenter trial which included 3,873 patients presenting with ACS undergoing PCI, patients were randomized to receive a fixed combination of clopidogrel 75mg and omeprazole 20mg versus clopidogrel 75mg and placebo [67]. Analysis of primary cardiovascular safety endpoints (death from cardiovascular causes, myocardial infarction, coronary revascularization, and ischemic stroke) did not reveal any difference between the placebo and omeprazole group with event rates of 4.8% with omeprazole and 5.7% with placebo (HR 0.99; 95% CI: 0.68 - 1.44) [68]. In summary, although pharmacodynamic studies suggest an existing interaction between clopidogrel and PPIs, the clinical significance is not as clear, particularly given the reassuring results of this randomized controlled trial. Thus for patients with clinical findings concerning for acid related disorders and those who require chemoprophylaxis, it remains reasonable to treat with PPIs to reduce gastrointestinal bleeding as the benefit of PPI outweigh the risk of cardiovascular event.

Conclusion

PPIs have drastically changed, and nearly revolutionized, the treatment of upper GI tract disorders. As one of the most commonly prescribed medications, it is important to recognize adverse events associated with chronic use. However, controlled trials establishing a cause and effect relationship between PPIs and several reported effects are sparse and therefore should be interpreted carefully. There is good evidence to support a relationship between PPI use and magnesium and calcium malabsorption, but less so for iron and vitamin B12 malabsorption. With regards to enteric infections, data suggests an increased risk of CDI and bacterial enterocolitis, but is questionable for SIBO. Furthermore, PPI use of less than 30 days has been associated with an increased risk of CAP, but this risk is not demonstrated in long-term use. Drug-drug interactions have been documented in multiple trials due to the effects of PPIs on the CYP 450 system.

In the appropriate setting, PPIs are safe and highly efficacious for the treatment of acid-related diseases. Short-term PPI treatment at standard doses does not represent significant risk. However, special attention should be paid to long-term PPI prescriptions particularly in elderly patients on multiple medications. In light of the current data, when the risk of PPI therapy outweighs the benefit, it is appropriate to withdraw or titrate to the lowest effect dose as clinically indicated.

References

[1] Heidelbaugh JJ, Metz DC, Yang YX. Proton pump inhibitors: are they overutilised in clinical practice and do they pose significant risk? *Int J Clin Pract.* 2012;66(6):582-591.

[2] Shin JM, Munson K, Vagin O, Sachs G. The gastric HK-ATPase: structure, function, and inhibition. *Pflugers Arch.* 2009;457(3):609-622.

[3] Wilhelm SM, Rjater RG, Kale-Pradhan PB. Perils and pitfalls of long-term effects of proton pump inhibitors. *Expert Rev Clin Pharmacol.* 2013;6(4):443-451.

[4] Reimer C. Safety of long-term PPI therapy. *Best Pract Res Clin Gastroenterol.* 2013;27(3):443-454.

[5] Lødrup AB, Reimer C, Bytzer P. Systematic review: symptoms of rebound acid hypersecretion following proton pump inhibitor treatment. *Scand J Gastroenterol.* 2013;48(5):515-522.

[6] Krupa LZ, Fellows IW. Lansoprazole-induced hypomagnesaemia. *BMJ Case Rep.* 2014;2014.

[7] Li M, Jiang J, Yue L. Functional characterization of homo- and heteromeric channel kinases TRPM6 and TRPM7. *J Gen Physiol.* 2006;127(5):525-537.

[8] Lameris AL, Hess MW, van Kruijsbergen I, Hoenderop JG, Bindels RJ. Omeprazole enhances the colonic expression of the Mg(2+) transporter TRPM6. *Pflugers Arch.* 2013;465(11):1613-1620.

[9] Thongon N, Krishnamra N. Omeprazole decreases magnesium transport across Caco-2 monolayers. *World J Gastroenterol.* 2011;17(12):1574-1583.

[10] Corleto VD, Festa S, Di Giulio E, Annibale B. Proton pump inhibitor therapy and potential long-term harm. *Curr Opin Endocrinol Diabetes Obes.* 2014;21(1):3-8.

[11] Wright MJ, Proctor DD, Insogna KL, Kerstetter JE. Proton pump-inhibiting drugs, calcium homeostasis, and bone health. *Nutr Rev.* 2008;66(2):103-108.

[12] Hansen KE, Jones AN, Lindstrom MJ, et al. Do proton pump inhibitors decrease calcium absorption? *J Bone Miner Res.* 2010;25(12):2786-2795.

[13] Chen J, Yuan YC, Leontiadis GI, Howden CW. Recent safety concerns with proton pump inhibitors. *J Clin Gastroenterol.* 2012;46(2):93-114.

[14] Ngamruengphong S, Leontiadis GI, Radhi S, Dentino A, Nugent K. Proton pump inhibitors and risk of fracture: a systematic review and meta-analysis of observational studies. *Am J Gastroenterol.* 2011;106(7):1209-1218; quiz 1219.

[15] Solomon DH, Diem SJ, Ruppert K, et al. Bone Mineral Density Changes Among Women Initiating Proton Pump Inhibitors or H2 Receptor Antagonists: A SWAN Cohort Study. *J Bone Miner Res.* 2015;30(2):232-239.

[16] Lee J, Youn K, Choi NK, et al. A population-based case-control study: proton pump inhibition and risk of hip fracture by use of bisphosphonate. *J Gastroenterol.* 2013;48(9):1016-1022.

[17] Valuck RJ, Ruscin JM. A case-control study on adverse effects: H2 blocker or proton pump inhibitor use and risk of vitamin B12 deficiency in older adults. *J Clin Epidemiol.* 2004;57(4):422-428.

[18] den Elzen WP, Groeneveld Y, de Ruijter W, et al. Long-term use of proton pump inhibitors and vitamin B12 status in elderly individuals. *Aliment Pharmacol Ther.* 2008;27(6):491-497.

[19] Lam JR, Schneider JL, Zhao W, Corley DA. Proton pump inhibitor and histamine 2 receptor antagonist use and vitamin B12 deficiency. *JAMA.* 2013;310(22):2435-2442.

[20] Skikne BS, Lynch SR, Cook JD. Role of gastric acid in food iron absorption. *Gastroenterology.* 1981;81(6):1068-1071.

[21] Sarzynski E, Puttarajappa C, Xie Y, Grover M, Laird-Fick H. Association between proton pump inhibitor use and anemia: a retrospective cohort study. *Dig Dis Sci.* 2011;56(8):2349-2353.

[22] Hutchinson C, Geissler CA, Powell JJ, Bomford A. Proton pump inhibitors suppress absorption of dietary non-haem iron in hereditary haemochromatosis. *Gut.* 2007;56(9):1291-1295.

[23] Stewart CA, Termanini B, Sutliff VE, et al. Iron absorption in patients with Zollinger-Ellison syndrome treated with long-term gastric acid antisecretory therapy. *Aliment Pharmacol Ther.* 1998;12(1):83-98.

[24] Howell MD, Novack V, Grgurich P, et al. Iatrogenic gastric acid suppression and the risk of nosocomial Clostridium difficile infection. *Arch Intern Med.* 2010;170(9):784-790.

[25] Deshpande A, Pant C, Pasupuleti V, et al. Association between proton pump inhibitor therapy and Clostridium difficile infection in a meta-analysis. *Clin Gastroenterol Hepatol.* 2012;10(3):225-233.

[26] Kwok CS, Arthur AK, Anibueze CI, Singh S, Cavallazzi R, Loke YK. Risk of Clostridium difficile infection with acid suppressing drugs and antibiotics: meta-analysis. *Am J Gastroenterol.* 2012;107(7):1011-1019.

[27] Janarthanan S, Ditah I, Adler DG, Ehrinpreis MN. Clostridium difficile-associated diarrhea and proton pump inhibitor therapy: a meta-analysis. *Am J Gastroenterol.* 2012;107(7):1001-1010.

[28] Seto CT, Jeraldo P, Orenstein R, Chia N, DiBaise JK. Prolonged use of a proton pump inhibitor reduces microbial diversity: implications for Clostridium difficile susceptibility. *Microbiome.* 2014;2:42.

[29] Sharma BK, Santana IA, Wood EC, et al. Intragastric bacterial activity and nitrosation before, during, and after treatment with omeprazole. *Br Med J (Clin Res Ed).* 1984;289(6447):717-719.

[30] Garcia Rodríguez LA, Ruigómez A. Gastric acid, acid-suppressing drugs, and bacterial gastroenteritis: how much of a risk? *Epidemiology.* 1997;8(5):571-574.

[31] Doorduyn Y, Van Pelt W, Siezen CL, et al. Novel insight in the association between salmonellosis or campylobacteriosis and chronic illness, and the role of host genetics in susceptibility to these diseases. *Epidemiol Infect.* 2008;136(9):1225-1234.

[32] Leonard J, Marshall JK, Moayyedi P. Systematic review of the risk of enteric infection in patients taking acid suppression. *Am J Gastroenterol.* 2007;102(9):2047-2056; quiz 2057.

[33] Evans CA, Gilman RH, Rabbani GH, Salazar G, Ali A. Gastric acid secretion and enteric infection in Bangladesh. *Trans R Soc Trop Med Hyg.* 1997;91(6):681-685.

[34] Ho JL, Shands KN, Friedland G, Eckind P, Fraser DW. An outbreak of type 4b Listeria monocytogenes infection involving patients from eight Boston hospitals. *Arch Intern Med.* 1986;146(3):520-524.

[35] Larner AJ, Hamilton MI. Review article: infective complications of therapeutic gastric acid inhibition. *Aliment Pharmacol Ther.* 1994;8(6):579-584.

[36] Cook GC. Infective gastroenteritis and its relationship to reduced gastric acidity. *Scand J Gastroenterol Suppl.* 1985;111:17-23.

[37] Cadranel JF, Eugene C. Another example of Strongyloides stercoralis infection associated with cimetidine in an immunosuppressed patient. *Gut.* 1986;27(10):1229.
[38] Campbell MS, Obstein K, Reddy KR, Yang YX. Association between proton pump inhibitor use and spontaneous bacterial peritonitis. *Dig Dis Sci.* 2008;53(2):394-398.
[39] Deshpande A, Pasupuleti V, Thota P, et al. Acid-suppressive therapy is associated with spontaneous bacterial peritonitis in cirrhotic patients: a meta-analysis. *J Gastroenterol Hepatol.* 2013;28(2):235-242.
[40] Lo WK, Chan WW. Proton pump inhibitor use and the risk of small intestinal bacterial overgrowth: a meta-analysis. *Clin Gastroenterol Hepatol.* 2013;11(5):483-490.
[41] Lombardo L, Foti M, Ruggia O, Chiecchio A. Increased incidence of small intestinal bacterial overgrowth during proton pump inhibitor therapy. *Clin Gastroenterol Hepatol.* 2010;8(6):504-508.
[42] Compare D, Pica L, Rocco A, et al. Effects of long-term PPI treatment on producing bowel symptoms and SIBO. *Eur J Clin Invest.* 2011;41(4):380-386.
[43] Choung RS, Ruff KC, Malhotra A, et al. Clinical predictors of small intestinal bacterial overgrowth by duodenal aspirate culture. *Aliment Pharmacol Ther.* 2011;33(9):1059-1067.
[44] Ratuapli SK, Ellington TG, O'Neill MT, et al. Proton pump inhibitor therapy use does not predispose to small intestinal bacterial overgrowth. *Am J Gastroenterol.* 2012;107(5):730-735.
[45] Sultan N, Nazareno J, Gregor J. Association between proton pump inhibitors and respiratory infections: a systematic review and meta-analysis of clinical trials. *Can J Gastroenterol.* 2008;22(9):761-766.
[46] Estborn L, Joelson S. Occurrence of community-acquired respiratory tract infection in patients receiving esomeprazole: retrospective analysis of adverse events in 31 clinical trials. *Drug Saf.* 2008;31(7):627-636.
[47] Laheij RJ, Sturkenboom MC, Hassing RJ, Dieleman J, Stricker BH, Jansen JB. Risk of community-acquired pneumonia and use of gastric acid-suppressive drugs. *JAMA.* 2004;292(16):1955-1960.
[48] Johnstone J, Nerenberg K, Loeb M. Meta-analysis: proton pump inhibitor use and the risk of community-acquired pneumonia. *Aliment Pharmacol Ther.* 2010;31(11):1165-1177.
[49] Giuliano C, Wilhelm SM, Kale-Pradhan PB. Are proton pump inhibitors associated with the development of community-acquired pneumonia? A meta-analysis. *Expert Rev Clin Pharmacol.* 2012;5(3):337-344.

[50] Canani RB, Cirillo P, Roggero P, et al. Therapy with gastric acidity inhibitors increases the risk of acute gastroenteritis and community-acquired pneumonia in children. *Pediatrics.* 2006;117(5):e817-820.

[51] Capurso G, Marignani M, Attilia F, et al. Lansoprazole-induced microscopic colitis: an increasing problem? Results of a prospecive case-series and systematic review of the literature. *Dig Liver Dis.* 2011;43(5):380-385.

[52] Keszthelyi D, Penders J, Masclee AA, Pierik M. Is microscopic colitis a drug-induced disease? *J Clin Gastroenterol.* 2012;46(10):811-822.

[53] Keszthelyi D, Jansen SV, Schouten GA, et al. Proton pump inhibitor use is associated with an increased risk for microscopic colitis: a case-control study. *Aliment Pharmacol Ther.* 2010;32(9):1124-1128.

[54] Beaugerie L, Pardi DS. Review article: drug-induced microscopic colitis - proposal for a scoring system and review of the literature. *Aliment Pharmacol Ther.* 2005;22(4):277-284.

[55] Muriithi AK, Leung N, Valeri AM, et al. Biopsy-proven acute interstitial nephritis, 1993-2011: a case series. *Am J Kidney Dis.* 2014;64(4):558-566.

[56] Ruffenach SJ, Siskind MS, Lien YH. Acute interstitial nephritis due to omeprazole. *Am J Med.* 1992;93:472-473.

[57] Sheen E, Triadafilopoulos G. Adverse effects of long-term proton pump inhibitor therapy. *Dig Dis Sci.* 2011;56(4):931-950.

[58] Simon T, Verstuyft C, Mary-Krause M, et al. Genetic determinants of response to clopidogrel and cardiovascular events. *N Engl J Med.* 2009;360(4):363-375.

[59] Lahner E, Annibale B, Delle Fave G. Systematic review: impaired drug absorption related to the co-administration of antisecretory therapy. *Aliment Pharmacol Ther.* 2009;29(12):1219-1229.

[60] Centanni M, Gargano L, Canettieri G, et al. Thyroxine in goiter, Helicobacter pylori infection, and chronic gastritis. *N Engl J Med.* 2006;354(17):1787-1795.

[61] Blume H, Donath F, Warnke A, Schug BS. Pharmacokinetic drug interaction profiles of proton pump inhibitors. *Drug Saf.* 2006;29(9):769-784.

[62] Wedemeyer RS, Blume H. Pharmacokinetic drug interaction profiles of proton pump inhibitors: an update. *Drug Saf.* 2014;37(4):201-211.

[63] Gilard M, Arnaud B, Le Gal G, Abgrall JF, Boschat J. Influence of omeprazol on the antiplatelet action of clopidogrel associated to aspirin. *J Thromb Haemost.* 2006;4(11):2508-2509.

[64] Gilard M, Arnaud B, Cornily JC, et al. Influence of omeprazole on the antiplatelet action of clopidogrel associated with aspirin: the randomized, double-blind OCLA (Omeprazole CLopidogrel Aspirin) study. *J Am Coll Cardiol.* 2008;51(3):256-260.

[65] Rassen JA, Choudhry NK, Avorn J, Schneeweiss S. Cardiovascular outcomes and mortality in patients using clopidogrel with proton pump inhibitors after percutaneous coronary intervention or acute coronary syndrome. *Circulation.* 2009;120(23):2322-2329.

[66] Ray WA, Murray KT, Griffin MR, Chung CP, Smalley WE, Hall K, Daugherty JR, Kaltenbach LA, Stein CM. Outcomes with concurrent use of clopidogrel and proton-pump inhibitors: a cohort study. *Ann Intern Med* 2010: 152: 337-345.

[67] Dunn SP, Macaulay TE, Brennan DM, et al. Baseline Proton Pump Inhibitor Use is Associated with Increased Cardiovascular Events With and Without the Use of Clopidogrel in the CREDO Trial. *Circulation.* 2008;118:1.

[68] Bhatt DL, Cryer BL, Contant CF, et al. Clopidogrel with or without omeprazole in coronary artery disease. *N Engl J Med.* 2010;363(20):1909-1917.

Index

A

acetylcholine, 3
achlorhydria, 21, 36
acid activation, vii, 1
acidity, 22, 74, 84, 103, 140, 142
acid-related disorders, vii, 2, 4, 11, 24
active transport, 2, 5, 7, 66
acute interstitial nephritis, 24, 38, 86, 105, 122, 142
acute kidney failure, 42
acute renal failure, 42, 104
adenosine, 2, 26
adhesion, 98
adjustment, 8, 11, 79
adults, 59, 103, 107, 119, 139
adverse effects, x, 10, 21, 125, 126, 139
adverse event, vii, viii, 13, 40, 77, 82, 107, 126, 135, 137, 141
African Americans, 10
aggregation, 136
allele, 75, 76, 77, 79, 80
amino acid(s), 5
anaerobic bacteria, 96
anemia, 65, 66, 83, 108, 130, 139
angina, 81
antibiotic, 99
anticoagulant, 13
anti-inflammatory drugs, 104
antipsychotic, 11
arrhythmias, 128
artery, 66, 75, 136
arthritis, 14
ascites, 37, 38, 85, 99, 100, 102, 120, 121, 132
aspirate, 96, 97, 120, 133, 141
aspiration, 89, 133
aspiration pneumonia, 89
asymmetry, 93
asymptomatic, 9, 104, 128
ataxia, 128
atherosclerosis, 56, 57
awareness, 100

B

bacteria, 10, 44, 98, 117, 132, 133
bacterial pathogens, 122
Bangladesh, 140
barriers, 92
bile, 30, 108, 116
bile acids, 116
bilirubin, 99
bioavailability, 8, 9, 27, 135
biopsy, 104, 135
biosynthesis, 9
bleeding, viii, 15, 16, 17, 25, 31, 32, 33, 39, 40, 41, 42, 44, 45, 47, 49, 65, 74, 99

blood flow, 40
blood pressure, 40
blood transfusion, 16
body mass index (BMI), 71
bone, viii, 22, 36, 37, 40, 46, 69, 70, 71, 72, 73, 110, 111, 112, 128, 129, 139
bone form, 70
bone marrow, viii, 40
bone mass, 111
bone resorption, 36, 69, 70, 73, 110, 111
bowel, 85, 97, 98, 119, 133, 141
breast cancer, 105

C

caffeine, 28
calcium, 23, 69, 70, 71, 73, 74, 84, 110, 111, 127, 128, 137, 139
calcium carbonate, 70, 110, 111
cancer, 24, 55, 123
CAP, 22, 23, 89, 90, 91, 133, 134, 137
carcinoid tumor, 24, 37
cardiac risk factors, 78
cardiogenic shock, 42
cardiovascular disease(s), 57, 136
cardiovascular risk, 46, 76, 77, 80, 116
cation, 109
Caucasians, 10, 76
causal relationship, 62, 96, 135
causality, 128, 135
causation, 22, 64, 83
CDAD, viii, 22, 40, 43
challenges, 96, 101
chelates, 108
chemical, 4
chirality, 27
chronic illness, 140
chronic kidney disease, 67, 89, 104
chronic obstructive pulmonary disease, 89
chronic renal failure, 59
cimetidine, 12, 18, 21, 34, 35, 50, 116, 130, 141
cirrhosis, 98, 99, 101, 102, 119, 121
citalopram, 87
classes, 53, 54, 56, 57, 62, 68
clients, 80
clinical application, 48
clinical examination, 53
clinical judgment, 94
clinical presentation, 104
clinical trials, viii, 30, 39, 62, 65, 134, 141
cloning, 25
clopidogrel, viii, 10, 11, 28, 29, 34, 40, 45, 46, 55, 58, 74, 75, 76, 77, 78, 79, 80, 81, 82, 84, 87, 88, 113, 114, 115, 116, 123, 136, 137, 142, 143
Clostridium difficile, viii, 22, 36, 40, 45, 46, 47, 48, 92, 117, 118, 119, 140
coagulopathy, 17, 41, 42
cobalamin, 107, 129
coffee, 8, 28
colitis, x, 40, 126, 134, 135, 142
collagen, 134
colon, 24
colon cancer, 24
colonization, 89, 92, 132, 133
colorectal cancer, 37
combination therapy, 11
commercial, 80
community, x, 22, 36, 50, 88, 89, 90, 116, 117, 126, 141, 142
comorbidity, 81
complications, vii, viii, ix, x, 9, 14, 18, 19, 20, 22, 40, 45, 46, 47, 61, 62, 64, 83, 106, 126, 140
composition, 131
compounds, 65
comprehension, 53
confounders, 65, 71, 86, 91, 97, 99, 102, 107
congestive heart failure, 89
consensus, 11, 29, 112, 136
consent, 54
constipation, 22
consumers, 56, 57
control group, 63
controlled studies, 128, 133
controlled trials, 15, 17, 18, 32, 47, 49, 87, 133, 137
controversial, 74, 85, 92, 98, 102

coronary artery disease, 29, 115, 136, 143
corticosteroids, 46, 71, 104
covalent bond, 2, 6
covering, 81
creatinine, 135
cross-sectional study, 63, 67, 109, 130
culture, 26, 96, 98, 120, 133, 141
cure, 10, 28
cyanocobalamin, 63, 107
cycles, 105, 106
cyclooxygenase, 13
cyclosporine, 67
cysteine, 6, 7, 9
cytochrome, 9, 10, 28, 75, 113, 114, 135, 136

D

data mining, 50
data set, 91
database, 63, 64, 71, 72
deaths, 81, 91
decontamination, 98
deficiency(s), 62, 63, 64, 65, 66, 83, 107, 127, 129, 130
delayed gastric emptying, 96
demographic characteristics, 53
demographic data, 94
Department of Defense, 125
detection, ix, 52
diabetes, 56, 57, 77
diagnostic criteria, 97
dialysis, 104, 110
diarrhea, viii, 13, 22, 23, 36, 40, 45, 46, 47, 48, 86, 97, 102, 103, 117, 118, 119, 122, 126, 133, 134, 140
diet, ix, 125
dilation, 20
discomfort, ix, 52, 90
diseases, 26, 27, 28, 119, 126, 137, 140
disorder, 20
diuretic, 66, 67, 68, 69, 83
diversity, 92, 118, 140
DNA, 79
dogs, 92, 118
DOI, 109
dosage, ix, 52, 53, 54, 56
dosing, viii, 2, 8, 11, 13, 15, 18, 20, 21, 24, 35, 64
double blind study, 50, 116
double-blind trial, 16
drug interaction, 9, 10, 24, 27, 29, 30, 106, 115, 123, 137, 142
drug metabolism, vii, ix, 10, 29, 61
drug therapy, 104
drugs, viii, ix, 4, 28, 30, 36, 40, 41, 43, 44, 45, 50, 52, 53, 56, 57, 58, 59, 84, 87, 88, 108, 110, 113, 118, 122, 135, 139, 140, 141
duodenal ulcer, 12, 14, 15, 30
duodenum, 6, 62, 64
dyspepsia, ix, 52, 53, 55, 126
dysphagia, 19

E

East Asia, 76, 114
ecology, 131
edema, 104
education, 58
elderly population, 63
electrolyte, 22
e-mail, 51
emergency, 67, 109
enantiomers, 6, 8
endocrine, 3
endoscopy, 55
energy, 2
enteric infections, x, 22, 86, 92, 102, 103, 121, 126, 132, 137
enteritis, 103
environment, 6, 8, 69, 70, 128, 129
environmental conditions, 96
enzyme(s), vii, 1, 2, 5, 6, 7, 8, 10, 45, 62, 75, 76, 84, 88, 113, 136
eosinophilia, 135
eosinophils, 135
epidemic, 126
epidemiology, 52, 119
epithelium, 35, 134

esophagitis, 18, 19, 20, 34, 59
esophagus, 112
ethnic groups, 136

F

family history, 80
fasting, 70
feces, 131
feedback inhibition, 4, 9
ferric ion, 130
ferritin, 130
filtration, 66
flatulence, 97, 126
fluoroquinolones, 45
fluoxetine, 87
food, 2, 8, 108, 139
Food and Drug Administration (FDA), 22, 24, 36, 37, 50, 62, 66, 69, 74, 105, 106, 108, 110, 126, 129, 131
force, 11
formation, 15, 65
fractures, 23, 37, 46, 69, 72, 73, 84, 110, 128
fragility, 71
freezing, 25
funding, 82
fusion, 2

G

gastrectomy, 21
gastric acid, vii, x, 1, 3, 4, 7, 8, 9, 17, 21, 22, 25, 26, 58, 65, 66, 69, 70, 83, 92, 98, 103, 110, 111, 116, 122, 126, 131, 135, 139, 140
gastric mucosa, 30, 48
gastric ulcer, 13, 14, 15, 30, 31, 35
gastrin, 3, 4, 9, 21, 25, 73, 127
gastrinoma, 20, 35
gastritis, ix, 22, 52, 55, 59, 142
gastroduodenal ulcers, viii, 2, 11, 12, 13, 14, 15, 16
gastroenteritis, 140, 142
gastroenterologist, ix, 52
gastroenterology, vii, 62, 67
gastroesophageal reflux, viii, x, 2, 11, 12, 26, 33, 59, 88, 116, 125, 126
gastroesophageal reflux disease, ix, 125
gastrointestinal bleeding, viii, 13, 15, 17, 27, 32, 33, 39, 42, 48, 137
gastrointestinal tract, 13, 23, 126, 133
generalizability, 84, 88
genetic factors, 106
genetic mutations, 68, 70
genetic predisposition, 134
genetics, 74, 140
genotype, 28
genre, 57
GERD, ix, 10, 11, 18, 19, 20, 21, 33, 34, 52, 82, 88, 96, 125, 126, 133
Germany, 39
gland, 2, 3
glucocorticoids, 42
glucose, 96, 133
goiter, 142
Gori, 114
growth, 97, 131
guidelines, 15, 17, 42, 43, 48, 59, 84, 85, 88, 95, 119, 136

H

H. pylori, viii, 2, 10, 12, 21, 57
H2RAs, vii, 1, 2, 4, 8, 12, 13, 14, 15, 17, 18, 19, 20, 21, 22, 23, 62, 64, 67, 72, 84, 86, 88, 89, 93, 94, 99, 100, 103, 104, 106, 127, 132, 133
half-life, 6, 8
HCl, vii, 1, 2, 26
head injury, 48
head trauma, 42
headache, 22, 44, 126
healing, 11, 12, 13, 16, 18, 19, 20, 21, 26, 31, 33, 34, 35
health, 23, 71, 78, 79, 80, 81, 89, 92, 122, 139
health care, 79
health care system, 79

health insurance, 80, 81
health status, 81
heartburn, 18, 34
Helicobacter pylori, 28, 63, 126, 142
hematocrit, 65
hematuria, 104
heme, 130
hemochromatosis, 130
hemodialysis, 104, 111
hemoglobin, 65, 66, 83, 130
hemorrhage, 15, 16, 31, 32
hemostasis, 17
hepatic failure, 44
hepatitis, 11
hepatocellular carcinoma, 101, 102
heterogeneity, 22, 68, 71, 72, 76, 77, 79, 82, 83, 84, 85, 87, 88, 89, 90, 91, 92, 93, 94, 95, 97, 103, 107, 117, 134
hiatal hernia, ix, 125
hip fractures, 24, 37, 72
histamine, vii, viii, 1, 3, 4, 21, 25, 31, 33, 35, 37, 39, 47, 49, 62, 73, 84, 108, 112, 139
histamine-2 receptor antagonists, vii, 1, 37, 47, 62
histology, 25, 55
history, viii, 13, 14, 20, 42, 52, 53, 71, 74, 128
homeostasis, 69, 70, 73, 109, 111, 139
homocysteine, 63
hospital acquired pneumonia, 91
hospitalization, 10, 23, 67, 68, 78, 79, 90, 104, 128, 136
host, 98, 107, 140
human, 2, 24, 25, 28, 29, 50, 92, 113, 128
human body, 2
human neutrophils, 50
human subjects, 92
hydrochloric acid, vii, 1, 108
hydrogen, 96, 129, 133
hyperparathyroidism, 69, 70, 73, 84
hyperphosphatemia, 111
hyperplasia, 37, 73
hypersensitivity, 104
hypertension, 56, 57
hypomagnesemia, 23, 67, 68, 83, 128
hypotension, 40, 42
hypothesis, 47, 72, 73, 92

I

ICU, vii, viii, 17, 39, 40, 41, 42, 43, 44, 45, 46, 47, 48, 49
identification, 78
idiosyncratic, 67, 86, 105, 106
ileum, 62, 129
imbalances, 87
in vitro, 10, 29, 46, 50, 75, 106, 110, 122, 131, 136
in vivo, 10, 27, 29, 122
incidence, 17, 40, 41, 59, 63, 94, 96, 101, 119, 131, 133, 134, 141
India, 104
individuals, 9, 64, 108, 111, 130, 139
infarction, 137
infection, 10, 12, 22, 28, 35, 36, 55, 85, 91, 92, 103, 104, 106, 117, 118, 119, 122, 132, 140, 141, 142
infections, vii, ix, 22, 45, 61, 86, 103, 119, 132, 133, 134, 141
inflammation, 104
ingest, 17
ingestion, 2, 8, 128
inhibition, vii, viii, 2, 3, 4, 7, 8, 10, 22, 27, 28, 37, 45, 70, 72, 73, 75, 77, 79, 84, 88, 106, 108, 109, 112, 127, 138, 139, 140
inhibitor, 9, 25, 26, 27, 29, 31, 32, 35, 36, 37, 38, 41, 44, 50, 59, 75, 108, 109, 110, 111, 112, 113, 115, 116, 117, 118, 119, 120, 121, 122, 126, 138, 139, 140, 141, 142
initiation, 23, 66
injury, viii, 2, 11, 13, 14, 17, 18, 21, 41, 42, 104, 122, 135
intensive care unit, vii, viii, 27, 32, 33, 39, 41, 44, 47, 48, 68
interstitial nephritis, x, 104, 122, 126, 142
intervention, 15, 43, 75, 82, 114, 115, 123, 136, 143
intestine, 66, 70, 119

intracellular calcium, 3, 4
intravenously, 29
ionization, 128
ions, 2, 3, 7, 129
iron, 23, 36, 64, 65, 66, 83, 108, 127, 130, 137, 139
ischemia, 40
Israel, 120
issues, 17, 22, 23, 46, 59
Italy, 51, 53

K

K^+, vii, 1, 2, 4, 5, 6, 7, 8, 20, 24, 25, 26, 27, 69, 72, 110, 111, 126
kidney, 42, 67, 89, 104, 122, 135
kinetics, 82

L

lesions, 17, 40
Listeria monocytogenes, 132, 140
liver, viii, 33, 40, 42, 45, 86, 99, 100, 101, 102, 121
liver disease, 33, 86, 99, 100, 102, 121
liver transplant, 42
liver transplantation, 42
localization, 25
long-term users, ix, 52, 53
lower esophageal sphincter, ix, 125
lumen, 7, 92
lung cancer, 89
lymph, 98, 121, 132
lymph node, 98, 121, 132
lymphocytes, 134

M

magnesium, 23, 66, 67, 68, 69, 83, 108, 109, 110, 127, 137, 138
majority, ix, 20, 43, 52, 56, 61, 64, 85, 86, 98, 102, 128, 135
malabsorption, x, 22, 65, 126, 127, 128, 137
malaise, 104
malignancy, 46
man, 27, 122
management, vii, viii, 2, 4, 13, 15, 16, 18, 21, 35, 48, 57, 113, 119, 126
mass, 21, 71
MB, 27
measurements, 76
mechanical ventilation, viii, 17, 39, 41, 42, 46
median, ix, 21, 52, 54, 56, 67, 96, 99, 106, 133
Medicaid, 63
medical, viii, 15, 19, 32, 42, 45, 46, 48, 52, 53, 54, 56, 69, 85, 87, 90, 91, 92, 93, 96, 101, 106, 118, 126, 128
Medicare, 79, 80
medication, viii, ix, 9, 11, 17, 24, 27, 40, 52, 55, 56, 59, 62, 64, 72, 77, 86, 90, 92, 94, 100, 102, 103, 117, 128, 134, 135
medicine, vii, 62
mellitus, 77
membrane permeability, 134
metabolic intermediates, 63
metabolic syndrome, 119
metabolism, 9, 10, 18, 27, 28, 29, 69, 73, 74, 75, 76, 77, 84, 88, 101, 113, 116, 135
metabolized, 9, 75, 136
mice, 4, 25, 26
microbial community, 131
microbiota, 92, 95, 96, 98, 108, 118, 119, 121, 131, 133, 134
microorganisms, 122
mineral absorption, vii, ix, 61
misuse, vii, 57
mitochondria, 2
models, 7, 24, 40, 70, 95, 128
molecules, 20, 98
morbidity, 21, 69, 133
mortality, viii, 15, 16, 17, 33, 39, 42, 43, 47, 48, 58, 69, 76, 77, 78, 81, 87, 91, 95, 100, 101, 102, 115, 116, 121, 133, 143
mortality rate, viii, 39, 42, 43, 91
motif, 5, 9
mucosa, 40, 43, 134

multivariate analysis, 45, 66, 71, 73, 81, 94, 95, 99, 101, 102
multivariate regression analyses, 46
musculoskeletal, 13
mutation(s), 10, 109
myocardial infarction, 80, 82, 113, 136, 137

N

nausea, 22, 44, 104, 126
nephropathy, 135
neurons, 3
neutral, 70, 74, 84
nitric oxide, 41, 48
nitrite, 48, 108
non-smokers, 24, 71
non-steroidal anti-inflammatory drugs, 42, 53
nosocomial pneumonia, viii, 40, 45, 46, 47
NSAIDs, 13, 14, 31, 55, 57, 90, 104
nutrient, 127
nutrition, 46
nutritional deficiencies, 126

O

obesity, ix, 80, 108, 119, 125, 126
oesophageal, ix, 25, 27, 33, 34, 35, 52, 53, 55, 57
old age, 66
organ, 2
organelle, 6
organism, 22
osteoarthritis, 91
osteoclastogenesis, 112
osteodystrophy, 23
osteoporosis, x, 37, 69, 70, 71, 74, 84, 110, 126, 129
outpatient(s), 56, 67, 69, 83

P

pain, ix, 21, 22, 44, 52, 56, 57, 78, 91, 97, 104, 126, 133
pancreas, 21
paracentesis, 99
parallel, 72, 82
parathyroid, 67
parathyroid hormone, 67
paresthesias, 128
parietal cells, vii, 1, 2, 3, 4, 8, 21, 25, 62, 112
paroxetine, 87
partial thromboplastin time, 17
participants, 65
pathogenesis, 24, 119
pathogens, 132
pathophysiological, 26, 69, 73, 84, 85, 98, 99
pathophysiology, vii, viii, 2, 22
pathways, 3, 4, 25, 45, 73
PEP, 57
pepsin, 15, 17, 62, 129
peptic ulcer, ix, 12, 16, 21, 28, 31, 32, 42, 52, 53, 55, 74, 82, 126
peptic ulcer disease, 12, 16, 42, 82, 126
peptide, 3
peritonitis, 24, 37, 38, 96, 98, 119, 120, 121, 141
permeability, 103, 119, 132
permit, 23, 133
pH, 2, 6, 11, 15, 16, 17, 28, 32, 55, 57, 67, 70, 74, 84, 103, 127, 130, 133
phagocytosis, 45, 50
pharmaceutical(s), 62, 90
pharmacokinetics, 27, 28, 29, 74, 84, 87, 88, 106, 114
phenytoin, 11, 29
phlebotomy, 130
phosphate, 110
physical activity, 71
physicians, viii, 2, 11, 20, 86, 106
Physiological, 26
physiology, 25
placebo, 11, 14, 15, 19, 28, 75, 82, 127, 136, 137
plasma levels, 10, 110
plasma membrane, 25
platelet aggregation, 75, 79

platelet count, 17
platelets, 10
pneumonia, x, 18, 22, 23, 36, 40, 43, 45, 46, 47, 49, 50, 85, 88, 89, 90, 91, 92, 116, 117, 126, 134, 141, 142
policy, 57, 125
polymorphism(s), 10, 28, 76, 84, 88, 106, 114
polyps, 24, 38
population, vii, ix, 10, 18, 29, 33, 36, 37, 52, 53, 56, 57, 59, 63, 64, 65, 68, 69, 71, 72, 73, 79, 81, 82, 83, 84, 85, 86, 89, 90, 94, 95, 100, 102, 107, 112, 114, 116, 118, 125, 132, 134, 139
pregnancy, 8, 13
prevention, 12, 13, 14, 17, 25, 31, 32, 49, 119, 122
principles, vii, viii, 2, 8
prodrugs, vii, 1, 5, 6
prognosis, 46
proliferation, 35, 98, 103, 131, 132
prophylactic, 17, 32, 90
prophylaxis, viii, 14, 17, 23, 27, 32, 33, 39, 40, 42, 43, 44, 45, 46, 47, 48, 49, 90
prostaglandins, 30
protection, 17
protective role, 47
proteins, 62, 64, 83, 105
proteinuria, 104
prothrombin, 17
protons, 67
prototypes, 26
PTT, 17
pumps, 8, 70
pyrosis, 56, 57

R

rash, 135
reactions, 22, 27
reactive oxygen, 50, 98, 121
reactivity, 6, 75, 136
reagents, 27
receptors, 3, 4, 73, 112, 121
recognition, vii, 1, 104
recommendations, 23, 42
recovery, 104, 135
recurrence, 34, 67, 94, 95, 127
reflux esophagitis, 19, 34
regression, 93
reintroduction, ix, 52
relevance, 7, 9, 79
relief, 9, 18, 19, 21, 33, 34, 53, 57, 102
remission, 13, 20
renal failure, 77
renal replacement therapy, 46
repair, 128
requirements, 16, 130
researchers, 90
residues, 6
resistance, 28, 70, 105, 122
resolution, 13, 18, 19, 21, 34, 67, 128
resource utilization, 17
response, 18, 28, 29, 75, 114, 142
RH, 33, 140
rheumatoid arthritis, 91
rings, 4
risk factors, viii, x, 17, 32, 39, 41, 42, 46, 47, 66, 71, 74, 83, 84, 85, 92, 95, 97, 99, 101, 119, 122, 125, 129
risk profile, 87

S

safety, 22, 24, 34, 35, 47, 69, 82, 105, 137, 139
saliva, 57
Salmonella, 103, 122, 132
salts, 30, 69
secretion, vii, x, 1, 2, 3, 4, 5, 6, 8, 9, 11, 17, 21, 24, 25, 26, 27, 28, 58, 70, 108, 110, 111, 116, 126, 127, 129, 140
self-control, 87, 98
sensitivity, 79
sepsis, 40, 42, 48, 49
septic shock, 42, 48
serology, 55
serum, 11, 63, 64, 67, 73, 83, 101, 109, 110, 135
sex, 42, 46, 66, 130

shock, 42
showing, 31, 65, 68, 75, 76, 83, 89, 96
side effects, viii, 22, 44, 46, 52, 53, 126, 136
signs, 86, 90, 106
simulation, 109
skin, 91
small intestine, 62, 70, 95, 98, 132
smoking, 71, 80
social costs, ix, 52
society, 11, 119
socioeconomic status, 63
sodium, 101
solubility, 11
South Korea, 99
species, 4, 103
spectrophotometry, 48
sphincter, 28
spine, 23, 37, 110, 129
spore, 22
SRMD, viii, 39, 40, 41, 42, 43, 44, 47
SS, 59
stability, 6
stabilization, 17
state(s), 2, 8, 12, 17, 21, 65, 69, 70, 79, 83, 113, 130
stent, 75, 76, 136
steroids, 135
stimulation, 3, 4, 8
stimulus, 2
stomach, vii, 1, 2, 25, 36, 37, 41, 98, 127, 129
stratification, 42
stress, viii, 2, 11, 17, 26, 32, 33, 39, 40, 41, 42, 43, 44, 45, 46, 47, 48, 49, 90
Stress-related mucosal disease, viii, 32, 39, 40, 47, 48
strictures, 13, 19
stroke, 80, 81, 137
structure, 4, 7, 25, 111, 134, 138
subgroups, 41, 94, 100
substitutions, 4, 5
sulfonamide, 7
Sun, 36, 117
supplementation, 24, 63, 64, 65, 74, 83, 84
suppression, 8, 12, 20, 21, 22, 23, 26, 35, 43, 44, 47, 63, 85, 92, 94, 95, 99, 103, 118, 120, 127, 128, 131, 132, 134, 140
surface area, 2
surgical intervention, 15
survival, 100, 103, 122, 132, 133
susceptibility, 103, 118, 122, 140
symptoms, ix, x, 8, 9, 14, 18, 20, 21, 27, 34, 52, 53, 57, 59, 82, 85, 90, 97, 98, 119, 125, 126, 127, 130, 133, 135, 138, 141
syndrome, viii, 2, 10, 11, 17, 20, 26, 29, 32, 74, 82, 114, 115, 131, 143
synthesis, 13, 40

T

target, 137
Task Force, 29, 113
testing, 55, 57
threats, viii, 40
thrombosis, 76, 91
tissue, 70
toxicity, 86, 101, 105, 106
transformation, 2
transfusion, 16
translocation, 98, 121, 132, 133
transmission, 22
transplant, 67, 109
transplant recipients, 67, 109
transport, 5, 105, 127, 138
trial, 9, 11, 14, 18, 19, 25, 28, 31, 34, 65, 72, 73, 75, 77, 79, 82, 97, 102, 110, 111, 114, 127, 136, 137
tumor, 20
turnover, 8, 70, 72

U

UGIE, ix, 52
ulcer, viii, 2, 11, 13, 14, 15, 16, 17, 24, 25, 26, 27, 31, 32, 33, 39, 40, 42, 43, 44, 45, 46, 47, 48, 49, 90, 126
United Kingdom (UK), 56, 72, 89, 112
unstable angina, 81

urea, 55
urinary tract, 91
urinary tract infection, 91
urine, 8, 73, 135

V

vagus, 3
vagus nerve, 3
variables, 77
variations, 87
vasopressor, 40
vector, 23
vein, 91
ventilation, 42, 46
vitamin B1, 23, 62, 63, 64, 83, 108, 127, 129, 130, 137, 139
vitamin B12, 23, 62, 63, 64, 83, 108, 127, 129, 130, 137, 139
vitamin B12 deficiency, 63, 64, 108, 127, 129, 130, 139
vitamin C, 65, 66, 83, 108
vitamins, 36, 127
vomiting, 22, 104
vulnerability, 41

W

water, 129
Western countries, 56
white blood cell count, 66
wild type, 10
Wisconsin, 125
withdrawal, 127, 128, 135
worldwide, 56, 105, 132

Y

yin, 11
young adults, 70

Z

Zollinger-Ellison syndrome, viii, 2, 11, 35, 131, 139